A Hacker, I Am

Craig Ford

Preface

I have considered writing a book on a few occasions but, to be honest, I never actually thought I would do it. However, here I am - writing this book. Or, more specifically, bringing together previous articles and chapters I have written to provide an insight into my opinions on the cyber security industry, its good and bad traits, and any other cyber related content I have an opinion on (some that I perhaps should have kept to myself, but you don't get anywhere in life playing it safe).

So why have I decided to do this? Simple really - the same reason I started to write articles on cybersecurity in the first place - I am really passionate about this industry. I caught the cyber bug back in 2013, although I had been in the IT industry since 2000 (wow, that seems like such a long time ago now). I really had only minimal interaction with the cyber security/hacker community, except maybe the love for the movie Hackers (I still think it's an awesome movie – a bit corny but definitely worth a watch). But when I started my degree, one of my first units was incident response, which I have to admit was the start of my security addiction - I was hooked from that point forward. I just needed to know more and more, I just couldn't stop (and I probably never will).

However, as with all areas in life, cybersecurity has its problems. I made a choice at the beginning of 2018: I wasn't just going to sit back and do nothing. I want to try and make a difference in this world, starting with the cybersecurity industry. I want to try and tell it how it is, and write in a way that a wider audience can understand - not just the technical folk.

I want this book to be accessible to anyone with an interest in cyber security and help them to believe that they can make a difference in their circles. I will bring in some fresh content, mix in some ideas I have already written about before for CSO, and try to have some fun with it all along the way. I hope you all enjoy the book and I have a feeling this may not be my only one. Till next time...

Acknowledgements

I just want to mention that without the help of CSO (IDG Communications) I would never have become the writer I am today. CSO has helped mould me into a good all-round writer for both this book and for the CSO website - in which I regularly write articles for, and was the original location for many of these chapters. It has been an interesting journey over the last year, and I believe that it has made me a more confident and well-rounded professional who is not afraid to give his opinion - even if it is a little out there sometimes.

I would like to also thank Jose Herrera Velasquez for working hard on the awesome illustrations in this book, and for helping bring to life the cover it bears. I was amazed at how dedicated he was to the task, and I feel that these illustrations have brought this book of mine to the next level. Considering I just gave him basic thoughts on what I wanted for each chapter, and his ability to put together an amazing set of illustrations was truly awesome work and I wanted to thank him for all his efforts.

A Hacker, I Am

CONTENTS

Chapter 1 - Managed service providers: the new target for cybercriminals

During 2017/2018 I noticed a common thread of discussion among my peers in the IT and Security industry, in which third-party organisations are the targets of malicious cybercriminals. Once these targets have been breached they are then used to gain access to larger and more valuable targets.

So why are they targeting third-party organisations instead of the primary targets? This is simple really - it's because these companies are generally small businesses who don't have the budget or means to implement sophisticated cyber security systems and procedures, making them an easier target than a cashed-up enterprise that has a well-established cybersecurity team.

This method for a malicious actor can be considered extremely fruitful, but when we consider the third-party organisation as a managed service provider or managed security service provider, these organisations have the "keys to the kingdom" in many organisations; with full administrator privileges and unlimited access to all systems, and minimal checks or procedures to minimise the actual access. If a cybercriminal gains access to the managed service providers/IT service providers systems, and can remain undetected for even a short period of time, that malicious actor could gain access to hundreds or even thousands of other organisations that the service provider supports.

Let's just consider this fact for a moment longer, using the attacker's perspective. One faces a potentially difficult target by pursuing an MSSP/MSP that could take a lot of resources to breach. But if access is gained with an unrestricted account they could gain access to every organisation that that business supports. Just imagine the potential for the malicious actor and the potential value this type of access could provide. The damage this type of breach could reap literally blows my mind. It would almost certainly cause the MSSP/MSP to go out of business, as it would be quite difficult to recover from an incident of this magnitude once it becomes public knowledge (which is inevitable, especially with an Australian organisation due to the recently implemented "mandatory breach notification laws").

This topic was brought up in an open forum/panel discussion at the ACSC cybersecurity conference in Canberra (April 2018 – The last ever ACSC conference – I didn't know that at the time though), and it was stated that the issues had been raised in the previous year's conference. However, I was very surprised at how the issue was widely acknowledged, yet nothing further had been done as an industry to try and reduce these risks. I am employed by an MSP/MSSP and the thought of a breach like this occurring at our organisation quite literally terrifies me. I immediately started to put together plans to sure up our protections, to ensure that we as an organisation do better with all aspects of security, and I honestly feel that we have made some great improvements.

I was also surprised to find that our organisation was just one of a very limited number who had sent any staff to this conference, which brings forward another problem. We have a known high risk of attack for MSP/MSSP, but very few organisations saw a benefit in sending staff to an event like the ACSC that could help us as an industry to collaborate and work out this problem together. Many of the attendees of the conference were government employees from various organisations, as well as large enterprise. No real representation of our industry, which is very disappointing, to say the least.

We need to come together on this initiative as an industry to share information on best practices, and lessons learned from both successes and failures. If we can share information more freely on threats or suspicious activity taking place through our own systems or our clients, we may be able to achieve a measure of success against a constantly climbing level of threat. I understand this chapter is a little strange - asking competitors to forget they're in competition with each other, and to then share information to help each other be more successful at protecting themselves - but it's about much more than just each individual business. Yes, it is about our clients and Australia. The better we all get at protecting ourselves from these types of threats, the better off we will all be. Whether it's the private sector or public sector we are all in this fight together.

Chapter 2 - Does your organisation have cyber insurance?

I am going to be bold here and speculate that the percentage of Australian businesses that currently have cyber insurance is very low. From the many conversations that I have had over the last few months it has become obvious that many businesses don't see any urgency or need for cyber insurance. Some just don't understand what it's really for, and why they need to have it.

With that in mind, I have decided to create this chapter to try and help readers from small businesses to enterprise organisations understand why they **need** cyber insurance. Let's start at the beginning, and outline a few statistics to indicate the cybersecurity problem as it exists in Australia. I will then outline what benefits cyber insurance can bring to your organisation. That way you can better make a decision on what cyber insurance packages are best suited for your organisation.

Without making this chapter a fear mongering exercise to terrify all business owners and readers, let's just look at some events that have occurred over the last year:

- Both Cadbury and TNT were both brought to a halt in June 2017 from a ransomware infection, with TNT appearing to be have been the most severely affected in Australia (at least from what has been made public anyway) with their parent company FedEx providing an indicative loss of $374 million from the incident. It was also indicated that several systems as of late 2017 were still not restored, and could be permanently lost. They had also indicated that operations had to be manually handled during the several months following the recovery, with some processes still being handled manually due to some systems never being fully restored.
- In October 2017, personal information of 5,000 Australian public servants of the Department of Finance, the Australian Electoral Commission, and National Disability Insurance Agency were publicly accessible because of a cloud service's mis-

configuration. There was also almost 50,000 private sector employee's personal information, which had been insecurely stored on an Amazon cloud storage service (just one of several worldwide over the last few months), and was easily accessible by anyone. This breach was caused by a private contractor who works with both government agencies and the private sector.

- On May 23rd 2018, PageUp - a hiring/recruitment software solutions provider - detected some unusual activity on its IT systems and publicly announced a possible breach on June 5th 2018. PageUp released the statement as required by the new data breach notification laws that had been introduced in February 2018. Will PageUp ever recover from this breach? Possibly not, due to the damage that it has suffered to its reputation and likely financial hardship it will face trying to rebuild trust among its customers.

These are just three of possibly hundreds of breaches that have occurred over the last year in Australia. It's hard to get an exact figure due to the mandatory notification laws only coming into effect in February. The reality is that cybercrime is estimated to cost Australian businesses of all sizes around $4.5 billion dollars every year, and there is evidence that this trend will only get worse as we become more and more reliant on data and our electronic devices for both personal/business use - not to mention that (almost) everything is interconnected via IOT.

So, what can you do to reduce your organisation's risks?

Some of these tactics are as simple as ensuring that you have an adequate set of policies/procedures in place, have your systems tested by a security professional, and train your staff to recognise phishing and scam emails. These strategies will help ensure that your systems are as secure as they can be, and that you're prepared to respond to an incident quickly and effectively when it happens. But what about the monetary costs involved with a breach?

The initial costs to a business caused by a security breach are easy to pinpoint, for example:

- Time lost to the organisation - from staff not being able to do their job, to labour costs for IT/security specialists to come in and recover your systems.
- Loss of income from not being able to access encrypted data for all outstanding invoices of which you don't have a physical printed copy. Some organisations will still pay, but you don't know what they owe, or if what they are saying is true when they say they don't have any outstanding invoices at all.
- Cost of new equipment and tools/software required to remediate or prevent a secondary incident occurring (it is always more expensive to secure systems after a breach rather than before an incident occurs).

So those are the basics that most people will be aware of. But what about the hidden costs?

- Loss of revenue due to the damage to your organisation's reputation.
- Your organisation could become the target of a lawsuit due to the loss of sensitive data. This would incur legal fees, and possible compensation payouts.
- The organisation could be fined for not meeting regulatory requirements, if this is something your organisation must adhere to.

The list could go on, but as you can see there are many costs relating to a breach that are not always obvious. This can help bring into focus the need to look at cyber insurance for your organisation.

What does cyber insurance cover?

Although policies will vary between insurers, a typical cyber insurance policy is designed to help you with both preventing breaches in the first place, and dealing with them if and when they occur.

Cyber insurance policies usually include the following:

- The cost of restoring or recreating electronic data following a breach or leak
- Forensic services to investigate a breach
- PR coaching in the event a breach harms your business's reputation
- Assistance guarding against data breaches, hacking, and employee error
- Guidance on how to respond to a breach
- Funds to cover the adverse financial effects related to a breach
- Funds to cover any fines that might be payable following a breach

Now you have the information regarding why you should consider cyber insurance, and what the policy will generally cover. It is very important that you clearly go through all of your options and understand any items that are covered, as well as situations/items that are not covered under the policy, as all policies are not equal. So, do yourself a favour and look into cyber insurance, so that when a breach occurs your organisation has the support it needs to survive. You will thank me later.

Chapter 3 - Penetration tests: What are the benefits? Should every business get one?

For many in the IT and Security industry, we are constantly bombarded with news articles and blog posts detailing the latest big cybersecurity breach that has just occurred - the avalanche of incidents never seems to stop.

It's no wonder that many business owners just don't know what to do, and are a little overwhelmed by the idea of trying to protect their systems. I feel that there are a few things that every business can do to protect themselves.

Most organisations invest in firewalls and antivirus protection - which is great as a starting point - but how can we protect a system adequately if we don't know what its weaknesses are? Do you know if your current protections will even help to stop an attack or ransomware infection spreading through your network?

The only way we can truly be confident that we have done the best we can do is to test our protections and make sure they react as needed. This type of test is typically known as a penetration test, and would be carried out by a penetration tester or security engineer who has the skill set required to conduct a thorough assessment and possible exploitation to conduct a true simulation of what a possible malicious actor would do.

The whole idea of penetration testing is a little strange, when you think about it. You're essentially paying a hacker/security engineer to break into your systems in anyway possible (obviously with some rules - we aren't criminals). Depending on the scope, the process could use social engineering, known (or even unknown) vulnerabilities, misconfigurations, or just exploit bad security practices (such as poor passwords, using the same password for every account, just to name a few). It is honestly quite scary how bad some security practices are until an incident occurs and businesses are forced to learn the hard way.

This is a process that every business should consider, due to the relatively small cost when compared to a real security breach, and the dramatic improvements/insight it could bring. Yes, on occasion no access point can be found, but there is almost always something that could be improved. Let's list some basic benefits of a security test:

- Find out any information that is publicly available from your organisation that can provide a cybercriminal attack vectors to manipulate.
- Discover unnecessary ports and services that are open to external access for no reason, or even just poorly configured systems that will allow an easy exploitation by a malicious actor.
- Identify poor patch management processes.
- Identify weak password or account management practises.
- Identify inadequate antivirus or perimeter protections.
- Plan a path forward for your organisation with clear recommendations on how to improve your organisation's security.

A pen test can take from as little as a week up to several months to complete - depending on many factors such as the size of the business, the scope of the tests agreed by the organisation, and what issues/vulnerabilities that were found during the initial reconnaissance phase of the test. This process and timeframes will be discussed by the organisation that you engage for the task, but it is always a good idea to have a clearly defined set of expectations and agreed processes that you are both happy with.

Chapter 4 - Could you spot a phishing email in their inbox?

Phishing campaigns have been a security concern for years - plaguing our email systems and clogging up resources. Yes, spam filters are getting better, but these pesky emails are still getting through to staff and are still claiming victims. Is this a lack of training? Or, are we going down the wrong track when trying to educate staff on what to look for?

I have been running user awareness-training sessions for clients over the last few years as part of my daily work (I work for Davichi an MSSP/MSP in Brisbane), and it is glaringly obvious that most people do not know even the basic identifying factors in a scam/phishing email.

In Australia so far this year there have been 10,862 victims of a phishing attack/scams reported to the ACCC (this number is likely to be much higher in reality, as this number is just reported incidents). The gender ratio for victims is evenly split between men and women with close to 50% on each side. It is also indicated by ACCC that $293,900 was lost as part of these scams, and only 1.4% of the people who reported the incident said they had a financial loss as a result.

Does this mean many of the victim's systems were infected by malicious applications/viruses for ransomware, or resources used for crypto mining? I would say that this might be worse than a direct financial loss if you take into account lost productivity, repair and clean-up costs, and just plain annoyance that this type of incident would produce.

For years I have heard people talking about user awareness training and making sure we all do it with our staff, but it seems to me that most organisations are conducting security awareness training solely for compliance purposes. Purchasing some online training package or a once off physical training session with staff, then ticking the box off for compliance to say that it has been done without giving it another thought. If that is how the organisations feel about the user awareness training value then it makes sense that most users don't find much value in the process.

It is my belief that we in the cybersecurity industry need to do everything we can to change this opinion for both business executives and all of their staff to ensure that every business gets the full potential of a security user awareness training program. Explain to them in plain English what we are trying to do and how it will not only help them but us all as a whole, and we will be on the right path.

Provide both educational and (if possible) slightly humorous training programs so that attendees remember what we have tried to teach them, don't over complicate it, and don't try to teach them too much in one go. Just make sure we teach them the basics, so they can better protect themselves online and during their normal business activities.

This is not a new concept I am talking about here, and is something that many of us are trying to achieve. But the part that surprises me is that most people don't do anything more than this. Do you simulate a phishing attack prior to the training session to gauge the level of understanding by your organisation staff? Do you publicly announce a leader or loser board of the results? Do you then follow up the training with another round of test/simulations? Or, do you just complete the training and have no further follow up?

User awareness training should have multiple methods, and should include a simulated phishing attack carried out against its staff. Users, in my opinion, should not be told of the initial phishing simulation so that we can get a true picture of how our users will respond to different types of email phishing campaigns.

These results should not be made public, and staff that fall victim to them should not be publicly shamed as this will only hinder the staff members learning. The information should be used by the team running the user awareness program to know what level staff are currently at, and to help gauge how effective the training program is at helping to improve the security awareness on follow up simulated phishing attacks.

If the results of the follow-up tests have no improvement or are worse than, our training program is not effective, and we would need to adjust our methods to try to ensure that we have a better result. This is one of the biggest problems I see in these types of initiatives/programs - they are run once by the organisation with no follow up testing to gauge if the staff actually improved, and the training programs are forgotten until the next compliance audit is coming up.

Do your organisation a favour and conduct a regular awareness program, and test your staff before and after the training programs to ensure you know if you are actually getting anywhere or not. If it isn't working, change things up, reach out to your peers in the industry and find out what works for them. You may be surprised how happy people are to discuss what they are trying to achieve and how they approach things.

We are all in this together, and everything is a work in progress. Keep trying to improve what your staff know, and they will come to see that you are just trying to help them. In the end, the better you can educate them, the easier your job will be.

Chapter 5 - Should small businesses care about cybersecurity?

There seems to be a constant bombardment of alerts landing in my inbox over the last few months about system breach after system breach. It really does seem like the future is going to be a challenging one for businesses when talking about protection from this ever-expanding avalanche of attacks.

Many of these breaches or cyber security incidents are being targeted at Australia's small businesses. This focus makes sense for several reasons - firstly, according to ABS, small businesses make up 97.4% of the total number of businesses in Australia. What do they classify as a small business? Small businesses are defined as 0-19 employees with a count of 2,066,523 small businesses. A further 2.4% are classed as a medium-sized business (20-199 employees), and a final 0.2% classed as large (200+ employees). That gives us an overall total of 2,121,235 businesses in Australia at the time of the ABS statistics.

With SMB's taking up 99.8% of all Australian businesses, it certainly makes sense that they are a target of cyber criminals. Let us look at the small businesses that make up this 97.4%; they have a small number of staff, a much smaller revenue stream than enterprise or even medium size organisations, and the pressure on their budgets would make it hard for them to consider looking at a really strong cybersecurity program (whether this is physical or software based protection, or even just user awareness training initiatives to help educate staff).

This makes these small businesses a great target for cybercriminals, because protections would be less, they would not have dedicated IT/Security staff to protect their systems, and if you really talk to these business owners they are usually very pressed for time. Cybersecurity is something that they may be aware of, but is right at the bottom of the list of their priorities.

Does that mean small business owners do not care about cyber security? No, I do not think it does, I feel that it is just put in the 'too hard' basket and left at the bottom of that list. Why is this so? I feel this is partially time and perceived costs, but to be honest I feel it is also our industry sectors fault. Many of the great protection platforms that are utilised by large organisations have a minimum requirement of at least 100 users/endpoints, and in some instances 1000 as a minimum user/ endpoint count. When small businesses are less than 20 employees that instantly rules these platforms out as possible solutions that they can use to help protect their systems and users.

Davichi - the MSP/MSSP I work for in Brisbane - is trying to resolve this problem for SMB's by adapting some of our solutions so that we can offer them to these types of businesses, but this problem will not be solved by one organisation. We as an industry need to work together at trying to help educate all Australian businesses (not just the big end of town) about how to reduce their risks of a cybersecurity incident, and how to better leverage options available to them.

I regularly talk to business owners in this market, and it is clear that this massive segment of small Australian businesses are not even considered by most of our industry due to the idea that it would not be cost effective to offer their platform/services to them. This view is very short-sighted, and we really need to come together as an industry in Australia - and around the world - to find a way to provide businesses of the smaller variety a way to utilise great cybersecurity solutions, because I really do not feel that they are leaving themselves poorly protected deliberately.

They only have access to basic antivirus solutions, and possibly spam filtering services (even some of these have 50 minimum user requirements), because that is all that we as an industry are making available for them. This just seems crazy to me - why can't we as an industry figure out how to make this work?

Let's come together and find a way to help protect these hard-working businesses. If we can better protect as many of them as possible it can only help make Australia as a whole more secure. Do not forget that many of our larger organisations use these small businesses as subcontractors, and that they may have access to your systems. Don't you feel that it is in our best interest to make them more secure?

Chapter 6 - Hackers are not the bad guys

Every day we are seeing more and more articles saying hackers have broken into another system somewhere in the world with an image depicting a person in a hoodie, usually in a darkened scene hunched over a laptop or computer screen, sometimes even with gloves on. Now, I want to point out some things that are very wrong with this.

Firstly, typing in gloves. That would be hard, and I feel that any self-respecting hacker would find it irritating at the very least to run their malicious code in the dark while trying to break into your systems wearing bulky gloves.

Let's break this down a bit. The hacker is hunched over the computer - which is certainly going to cause some back pain. The lights are off, meaning eye strain problems (we all know those pesky blue lights from the screens are damaging our eyes). The hoodie and gloves are on, so every time they try to run their commands they keep typing the wrong keys, and spend five times as long trying to execute their malware which they have bought on the dark web. Seems like a very stressful situation if you ask me.

All jokes aside, I have been hearing lots of chatter amongst my peers trying to convince the world that "hacker" is not the correct word to use when trying to describe a malicious actor breaking into a company's system, or infecting them with one of those awful crypto bugs. The word hacker means, "an enthusiastic and skilful computer programmer or user" (as far as the dictionary goes), but I personally class a hacker as someone that tries to find ways to access or use a program or system in a way that it was not intended to be used. To - yes, break things - but not always in a malicious way. It could be to test the strength of a security protection, or input/output manipulations just to see what they will do. Many of these people are the ones out there now protecting your businesses.

If you have been to any conference related to the security industry in the last 10 years, you will have seen thousands of hackers in one place. Ordinary looking people you would see in your street, or even your neighbours, could be put into the category of 'hacker'. I would personally classify myself as a hacker, probably not a very good one, but nonetheless I would fall into that category.

I work as a senior security engineer and I do penetration tests as a primary part of my daily duties. I try to manipulate and break our client's systems to just see if I can - but the difference is once I find a problem, I help the client fix it. I am a hacker.

I don't remember seeing a hooded, shady looking character at any of these security events over the last few years, trying to cover their face so that no one can identify them. Our team regularly attends these types of events to ensure we keep up with what is happening in our field. This 'hacker' image that is constantly portrayed just isn't realistic. Yes, it is a caricature that Hollywood has been using for years to portray a hacker, but just because everyone recognises it, doesn't mean it's accurate.

Let's look at the statement of a hacker - they are the cause of all incidents, right? Wrong. Not all hackers are underworld figures, cybercriminals, or just mischievous people who want to take you and your business for a ride and hold you for ransom over your own data. No; hackers are normal people with normal jobs that, in many cases, are trying to help businesses protect their systems from being breached by the real criminals or malicious actors - just like I do.

I used two words in the first half of that paragraph that could better describe someone who has hacked into your systems and encrypted your files. Let's list some other options here:

- Malicious actors/s
- Cybercriminal
- Criminal
- Cyber-terrorist
- State-sponsored cyber-attack groups

These are just the first five that I thought of, but they are already better than using the word 'hacker' when instead you mean 'criminal'. It's really that simple. If we go back to the security conferences that are full of people that look more like professional businessmen and women in suits or formal attire - *that* would be a better depiction of what most hackers look like. Yes, some of us like to wear jeans, and possibly some geeky t-shirts (I personally like stupid slogan t-shirts), but don't tarnish us all with the same brush. Most of us are actually here to help, not try to get your secrets and sell them to the highest bidder.

If you are writing a article or white paper consider what your discussion is about, and use the correct name to describe the culprits responsible for the incident or breach. The 'cybercriminals' or 'malicious actors' are my favourites - simple, but effective, realistic, true statement. Not 'hackers', which tarnishes us all with that criminal brush.

I feel if we can all do this, we will all be better for it.

Chapter 7 - Perfect security is a myth, stop selling a lie

Since I started writing security articles I have had quite a few people approach me asking about my opinion on different security solutions. I don't mind sharing my opinion on things, but they are just that - my opinion. I like what I like, and am very happy to burst the bubble of any security solution that tells people their solution will protect them from all threats and that nothing will get through.

This is the biggest load of crap, and yet I keep hearing it being spouted to customers by vendors and other security service providers. **STOP** telling potential customers that your new suite of cloud security products, or some fancy new physical firewall will protect them from all threats that may come their way.

It really gets under my skin when I hear a pitch being spun this way. **NO** systems or protections are 100% perfect and can defend your organisation from all threats 100% of the time. Yes, there are some great systems out there that can give you really great protection, but true security is about more than just buying the latest and greatest protection and saying, "I am protected now, nothing else needs to be done". Whoever is selling you that pipe dream is worse than a slimy car salesman.

Don't fall for the hype. Do your homework and talk to security professionals about what you should really do to protect your systems. Many will be happy to point you in the right direction. Not all will agree on the best solutions or practices that will work for you, as we are only humans after all. We all have favourites that we like to use more than others (sometimes for stupid reasons that only make sense to us). The point is, don't fall for the silver-tongued salesperson who is trying to sell you the cybersecurity equivalent of a stylish, new red convertible. *Trust me*, you will regret it in the long run.

There is no silver bullet solution that will protect you, but it does not mean you should stop trying to make your systems safer. Instead ensure that you get a solution that provides the features that work best for your environment. Make sure that you have a good set of security policies, train your staff so they know what to look out for, and make sure that they are comfortable with approaching your security/IT team if they are unsure about something. Chastising people or making fun of them is never helpful when trying to educate them. If you can make them feel as though you are here to help and have no judgement, they will come to you when they see something funny. This goes for the staff on your IT helpdesk as well. They are the people who see all the issues, and can hopefully give you a heads up that something is going on. This could help stop an incident or breach before it goes too far.

Patch your systems more than once a year. Please do this, such a simple thing can be what saves you. Yes, I know some of you may have systems that makes it hard to do regular updates, but just because it is hard doesn't mean you shouldn't do it as much as possible. Ensure that all systems are run through a hardening process before you put them into a production environment. Yes, I know you have deadlines and you don't really have time to do this. *Make time*, you will be grateful you did later. And don't just stop there.

Security is a work in progress. Keep working on making your systems more secure and you will be as prepared as you can be when an incident occurs.

Let's do a quick recap. Do not trust the silver-tongued salesperson selling you the silver bullet solution (that new red sports convertible) that will solve all of your organisation's security problems and enable you to reduce your security team. What they are selling you is pure science fiction. Get the option that best meets your needs, but do it with your eyes open and don't fall victim to the hype of the next best platform. They have just thought of a new way to sell you the same thing that all the other vendors have been providing for years.

Cover your basics and remember that security is always a work in progress. Keep making improvements and working towards the best security you can, and you will greatly reduce your risk of a breach. A final note on this - some breakthroughs really are breakthroughs. If the new solution really checks out and it will solve a problem in your systems, then go for it. But remember - *do your homework*.

Chapter 8 - You want a career in cybersecurity? Are you crazy?

So, you want a job in cybersecurity. You have seen the latest hacker movie, or Abby on NCIS, and want to save the world from all the malicious actors and cyber criminals that are out to get us? You have drunk the Hollywood Kool-Aid and have fallen for what a life as a white hat hacker saving the world is like, and you must be a part of it. What is it really like working in cybersecurity, though?

60-80-hour weeks, endless alerts and logs to review and determine an effective course of action, never-ending training requirements just to keep your head above water with the latest trends and attacks that are being used by malicious actors. More letters after your name than the alphabet just to ensure you mean something and can even get a look in for the most basic of security positions. And that's with a skills shortage - what will it be like when there are too many candidates flooding the job market? Although, I don't feel we will have that problem for a while.

Don't get me wrong, I love working in security, and I will outline the general path to get here. However, I want to ensure that anyone looking to make a shift over to security doesn't come into the industry with rose coloured glasses on.

I want you to see what it's really like, and then you can make an educated decision to join us. Let's start with how I got here. I have been in IT (Information Technology) since 2000 (after I left high school), and have been employed by several different organisations since then, in internal enterprise support roles, IT service provider support, and management roles.

I found myself in 2011/2012 seeing a lot of security-related incidents with viruses and other similar incidents. I got the security bug at this time, and started to learn as much as I could about anything security to improve my skills as an IT professional (well, that's what I thought at the time). I continued to improve my skills until 2013 when I started my first master's program with CSU and signed up for a digital forensics major. I have been focusing more and more on security as my primary duty since, with a change from an IT role into a security specific role at the start of 2017 *(best thing I have ever done)*.

My entry into the security specific role was not easy though, as I had wanted a change for several years but was looked over for security-specific positions because of my lack of security specific role experience, and I didn't have any of the most popular security certifications (CISSP, CISM, OSCP – the list goes on). I have now almost completed a second master's in information systems security with CSU, as I wanted to expand my qualifications further due to the resistance I experienced (even after my first degree). I am growing as a security professional every day, with a real passion to continue to learn and be an active member in our industry.

As a security professional I am still a newcomer to the industry when compared to many of my peers, but I feel that it is not always how long you have done something that makes you a positive participant, and I want anyone who is considering making a move into this industry to truly know that if you want it bad enough you can make it happen. It isn't going to be easy, and you are going to need to work hard to achieve the skill set needed to be a good security professional (I still have lots to learn and will probably never know everything I need to know), but start with the basics and build from there.

You can find many resources online to help with that. And don't be afraid to push yourself out of your comfort zone. If university is something that you would like to do, then find the course that will interest you the most. Passion will help you get through the tough times on that journey (and trust me, there will be some tough times – life has a habit of getting in the way). Whichever path you take, don't be disheartened by a rejection. These are inevitable, and they're just something that you will learn to turn into motivation.

The other fact that many of you will need to know is most of the time a job in cybersecurity is not flashy or exciting. Many jobs on the lower level in security are responding to alerts and reviewing log information. This will not be super exciting, in fact it can be quite mundane, but these tasks will make you a better security professional, because if you know how to find the abnormalities and how breaches are carried out you will greatly improve your skills. As you develop you can specialise, but I can't say it enough - always keep working on the foundational skills, you'll thank me later.

If continuous learning and mundane long hours is something that still interests you, then go for it and enter with your eyes open. Hopefully, we will get to work together someday in this truly interesting security industry.

A quick opinion for the people doing the hiring - if there is such a big shortage of skilled staff for advertised positions all around the world, maybe you should consider hiring people who have the basic skill set you need and train them. They will be grateful for the opportunity, and maybe turn out to be the strongest people you have on your team if you just give them the chance to show you what they are made of. If you invest some of your time in someone and they don't work out, have you really lost out on that much if you couldn't fill the position otherwise, and would still have been short staffed anyway? Just a thought!

Chapter 9 - FiveEyes are giving themselves a backdoor into your systems

The Five Eyes is an alliance comprising of Australia, Canada, New Zealand, the United Kingdom and the United States of America. The alliance was formed to share intelligence, especially signals intelligence, between countries via the UKUSA agreement. As the name would indicate, it was originally created in 1946 (originally called BRUSA). During the late 40's and 50's many countries under the UK's power started to exercise greater control over their own dominions, leading them to start to represent themselves in the intelligence sharing pact (Canada in 1948, Australia and New Zealand in 1956).

These countries are involved in other intelligence sharing agreements via organisations such as NATO, but much more information is openly shared via the Five Eyes agreement than any other. In a meeting at the end of August, the countries discussed a range of proposals to combat terrorism and crime, with an emphasis on the internet. They believe they should encourage technology providers to establish lawful access capabilities to their solutions to help intelligence and law enforcement organisation protect our countries. This access is mainly aimed at encrypted communication services like WhatsApp, Wickr, iMessage, and even Snapchat, just to mention a few.

The Five Eyes made a statement at the end of the event. A part of this was, "Should governments continue to encounter impediments to lawful access to information necessary to aid the protection of the citizens of our countries, we may pursue technological, enforcement, legislative or other measures to achieve lawful access solutions".

Soon after, Australia proposed new decryption laws that would allow just that - Australia's Assistance and Access Bill 2018. This bill was introduced to parliament on the 20th September 2018, and was referred to the Parliamentary Joint Committee on Intelligence and Security for inquiry and report. The Committee commenced its review and was accepting submissions until 12 October 2018. For more information please visit the Committee's website. The Bill as introduced into Parliament and the Explanatory Memorandum can be viewed on the Australian Parliament House website.

Now I understand why the government and the Five Eyes as a whole would want this type of law introduced, as a security professional trying to investigate incidents and knowing a lot of the techniques that forensic investigators would utilise to try and gain access to suspects data and communications to either prove or rebuke claims against someone. Encryption can be a massive roadblock. Not being able to see communications between suspects or view stored content can greatly hinder investigations. With technology advancements continuing to evolve, this will become a growing problem for law enforcement. But is this type of backdoor access they want a good idea?

Let's look at the two big concerns I have with this initially:

- Creating a backdoor in secure applications will undoubtedly provide unauthorised access to malicious actors - that is inevitable.
- Who will honestly and ethically manage the access to ensure that it is handled correctly, and that we as citizens are protected from unlawful abuse of powers? Who will restrict what organisations can even request access, and to what systems does the law allow this access too?

So, let's run through this. Software companies will create the access required by law, and a couple of weeks or months later a malicious actor has gained access to the back door and stolen sensitive data, which it will sell to highest bidder. This malicious actor could have several different targets, and this will just make their job easier. Why spend months trying to find a hole in an application or security systems, if someone is just going to hand you one that will be much more efficient and likely easier to gain access to? No matter how well these backdoors are protected, it will only be a matter of time before someone gains unauthorised access via this new 'golden ticket' into previously secure systems - obviously they were secure if legislating backdoor access is the only way for law enforcement to get access.

I think this is the likely result of any backdoor being created in applications, no matter if it is a result of this law being passed or not. This law is not the way to achieve the result they are after. I feel that they need to find a different solution that does not create a huge risk like leaving the back door open to our messaging apps or secure data storage.

So what if I'm completely wrong, and no one - not even North Korea or China - finds a way through the new wide open back door? How can we as a society manage access to these powers, without hindering the ability for law enforcement to get access to the data they need? And, access it fast enough so they can keep us safe from growing threats? But also minimise the access to this information for purposes not so honourable?

Will the new laws, if passed (I think it is inevitable but I will go with "if" anyway), make Australia an easy way for our partners in the Five Eyes to gain access to information that their own countries don't currently allow access to? And, will this type of law hinder future investments in new tech start-ups due to the concern consumers will have for their privacy? I feel some of our larger partners will certainly try to leverage or pressure their Australian counterparts to provide such access. Which brings us back to the previous statement - who will hold the preverbal keys to the kingdom (or backdoor in this case)?

I honestly don't have an answer for this, and really don't know how our government is going to approach this. I truly hope they have considered every aspect, and that protections are put in place that will ensure that minimal abuse of these powers is possible.

I guess only time will show how this is going to play out. If the proverbial kingdom burns down then we'll just bring the marshmallows and enjoy the show (while trying to resist the urge to say we told you so)...

I am now imagining the onslaught of people saying that, "If we don't have anything to hide, then why do we have an issue with these laws". But, this is my opinion, and I want to be clear - this isn't about privacy for me at all. It's about the possible abuse of power, and opening up a huge security hole in previously reasonably secure systems, which would provide access to people that these laws are meant to protect us from.

Chapter 10 - What to look for when hiring security talent: hidden talents

Cybersecurity and ICT security talent is a highly sought-after commodity in today's market. In Australia this seems to be driving up the cost for companies to actually obtain security talent. This is understandable, considering the increasing level of breaches we are seeing worldwide, and this isn't going to slow down anytime soon from what I can see.

So how can we find the talent we need, to fill the thousands of positions that we need to fill, in order to ensure that our systems are as protected as they can be?

Let's first look at a problem which I discussed in my previous chapter, "You want a career in cybersecurity? Are you crazy?". Basically, the problem is many hiring managers and recruitment professionals are overlooking great candidates for positions that they could shine in, for two reasons - one of which is a bit of a conundrum.

The first is due to candidates or budding cybersecurity professionals not having those coveted certificates such as CISSP or CISM. I understand for senior positions that this may be a requirement to solidify one candidate over another, but for entry-level security positions is this realistically something that organisations can ask for?

Let's review the CISSP Cert, just as an example, as it is a regular requirement in security positions. With the CISSP you need to study and take the exam (which costs around $700 to sit), but to be awarded the Cert you need to have 5 years of experience in one or more of the 8 security domains. If you don't have the required experience you are only able to be awarded an Associate of $(ISC)^2$. You will then have a maximum of 6 years to gain the 5 years of experience to be awarded the CISSP.

So, if the CISSP, for example, is a requirement of a junior security position, then how do you acquire the 5 years' experience to get the certification if they won't give you the position without it? Strange situation - need the Cert to get the job, but can't get the job without the Cert. I really don't know how that one works, but it doesn't seem ideal for the recruiter or the candidate.

That brings us to the second reason many are being knocked back for positions, and that is lack of experience. For some reason companies want someone with 10 years security experience to fill junior security jobs, when people with 10 years' experience are overqualified for these positions. This is the conundrum I mentioned earlier - companies want someone with experience to fill junior positions, that most applicants for these positions don't have, and the people who do have that level of experience wouldn't even consider the junior position that is being offered.

This is a vicious cycle that just won't solve our problems, especially if we all stay on this same path and don't look for candidates outside of the box. Let's dive into some things I personally feel companies should look out for that will give them a great candidate for a junior position. Not just industry Certs or specific experience, especially if you have already tried to get a candidate of that level and received little interest.

1. **Interest in Cybersecurity** – if a candidate has a strong interest in cybersecurity, and can show they have been trying to educate themselves, this to me shows that they will be a hard worker and really push to achieve the skills that are needed for the position.
2. **Programming or strong IT background** – if a candidate has a strong background in either, or both (both would be a unicorn – but it could happen), and they have that strong interest then they could become one of the best security professionals you have - with some time spent developing them.
3. **Industry certifications or formal training** – in this situation you may have a candidate who can show that they have spent the time to complete training in areas of security, and build those foundational skills that are needed to be able to successfully work as a junior in this industry.
4. **Participation in the security community** – if a candidate attends meetups or is an active member in industry associations, it shows that they are trying to get involved and become an active participant in our industry. Candidates can learn a lot from the people they meet at these events, and I feel that we could all learn

something from our peers no matter what our experience level is.

5. **Attitude and personality fit** – I find this one to be very important, and to me can be more valuable than experience or a certification in many situations. If someone has that right attitude, and will fit well with the current team, they will connect and learn from them quickly. They also won't cause issues in a team that works well, which can be a big problem in our industry, with ego's sometimes getting in the way.

Now, don't get me wrong, this is not an exhaustive list, and there are many more things that a candidate could need. But, I feel that if you find someone with these characteristics, and you have a team in place that could help bring them up to speed quickly you should consider helping that candidate out. If you give them a chance to get into the career they have been really trying to get into they will be a loyal member of your team, and not forget that help you gave them in the beginning.

I know this for a fact as I have been given similar opportunities in the past to prove myself in a position that I might not have been 100% qualified to do, and I believe that it has worked out well for both parties. I was a loyal employee and worked very hard for the organisations I have worked for. That doesn't mean you stay with them forever, or that every chance you take on someone will pay off, but if you truly give someone a chance it could be the best thing you do all year, or even all decade.

Some of you may just dismiss my opinions, but please truly consider what I have said, and don't just do what you have always done - give someone a chance. As I always say, if you don't agree with my opinion, that's fine. We are all individuals and have our own views, but we should keep an open mind and be open to others' views.

Chapter 11 - We don't need penetration testing, we're in the cloud

You have your servers hosted in the cloud on one of the latest - it's faster and more secure than any that have come before it. You do not need to have your virtual servers tested to ensure they're secure (or as secure as they can be), that's just a waste of your time and money.

The provider is responsible for ensuring that your systems are safe from any cybercriminals or script kiddies that want to bring your systems down just for the entertainment factor, or to encrypt all of your data, just so they can squeeze you for every penny you have. That is not our responsibility. We have antivirus on the servers to protect them, and I do not know what your testing could do to help improve anything.

This is a story I have heard on many occasions over my career in both IT and Security - that a customer has "the cloud", which I don't think many understand is just a physical server platform hosted in someone's data centre, or in offices that they rent out at a price that allows them to share the costs of the equipment between clients and also make some profit for them.

A cloud-hosted server is not hosted in some magical place that is impenetrable to all of those above cybercriminals - it is still on physical hardware, just like the one you would put at your own premises if you weren't in "The Cloud". They are still hosted on some form of Linux, VMware, Nutanix platform, and are just as vulnerable to threats as they would be in your own building. Yes, they are probably much more expensive servers than you would probably get, but that does not change the fact that they are still normal servers.

I want to clear some things up in this chapter about this belief. Providers are only responsible for all of the security on your cloud service if it is a hosted application. In this case, you still need to keep your passwords at correct levels to help keep things secure, but the platform it is hosted on is the provider's responsibility.

If you have virtual servers hosted on the cloud, the provider would only be responsible for the underlying hypervisor platform and the data-centre or site that it is hosted. You as a customer are still responsible for keeping your servers secure and protected from the operating system upwards.

If you have a physical server hosted in a data centre, then you are responsible for everything regarding security on those physical systems, and the cloud provider is only responsible for the physical data centres location and its networks.

It is also true that if you have hosted services in the cloud and decide that you are going to get a penetration test and/or a full security audit completed you need to notify the provider that this is going to be occurring, and have a clearly defined scope of what is going to be tested. Don't just go at the systems like the wild west with guns blazing (aggressive vulnerability scanners, brute force attacks, RDP attacks, systems exploits or whatever else the tester wants to throw at the systems). This will result in some very unhappy people when they find out who was responsible for the incident.

Even if you don't get an unhappy call regarding a test, it is really just the legal and ethical right thing to do. You can do anything you want with your own systems hosted internally, but cloud-hosted platforms are generally shared platforms. They could bring heavy legal penalties to the testers if they don't have a good scope outlined for the job and don't have the necessary authorisation to conduct malicious activities on a client's behalf.

Be smart. Get it all in writing and ensure that everyone, including the hosting provider, knows the periods for the tests, who, and what is going to be executed against the systems ahead of time.

Now that we have covered what you are responsible for, and what the service provider is responsible for, as well as the requirement for all parties to be aware of what is happening, we really need to cover why you should spend the budget on getting the tests done in the first instance (cloud or no cloud).

Do me a favour and look over some more chapters in this book, or open up a browser and type "cybersecurity breaches 2018". You will see so many chapters about breaches, and that is just for 2018.

Do not fall for this same load of crap; "I have insurance; I do not need to test my systems security, or make sure my users abide by our system usage policies, isn't that why we have the cyber insurance?". Look at some of those chapters which you would have just found on CSO, or in your Google search - they will nearly all depict the same scenario. Millions of dollars in damage files lost, companies going out of business because they have lost all of their data, and a new reputation that makes them seem worse than the plague to potential customers.

Do your organisations a favour - get your systems tested, get the best protection you can afford, ensure you have proven policies in place. Make sure you actually test incident response and data recovery plans. Don't just make them and forget about them, as you will regret it when an incident does occur (and trust me, it will). Train your staff, and yes, have cyber insurance. All these things combined will keep your business running and your reputation intact.

It is much better to prevent an incident then try to clean up after it.

Chapter 12 - Autonomous cars are coming, and hackers are rubbing their hands in anticipation

Over the last few months I have been doing lots of research about new cars - particularly family sized cars that can handle a growing family for many years to come, as our current family car is getting a bit old and could do with an upgrade. As a security professional (would be hacker) I seem to really pay attention when the car salespeople start to talk about all the new connected features - like the auto link app in the new Hyundai Santa Fe, that can record all the details on where the vehicle has been. Some other features I've heard include:

- Owner driving analysis (driving habits, scores)
- Alerts roadside assistance in car breakdown
- Sends emergency alert messages upon a crash
- Logs private and business trip stats for record keeping
- Remote access to doors and climate control
- Provides fuel level and efficiency
- Allows remote engine start from a smartphone
- Provides service bookings at your fingertips
- Tyre pressure and battery status
- Tracks driving and refuelling history

I'm not sure that I want some of these features, but I guess if someone has an accident in the vehicle and is unconscious it would be comforting to know that help is on the way without any further interaction from the occupants of the car. The tyre, battery, and servicing abilities are a nice touch. Some of the remaining options, however, could be quite entertaining. If I was to gain access to a person's account, I could turn up the heating in summer as high as it goes, or turn the air conditioning up so that it is like driving in a snow storm. But these are just some silly pranks you could play on an unsuspecting driver (or my wife, in my case).

What about the remote engine start and remote unlocking of the car? The only real reason that you would want to remotely start a car is to turn on the air conditioning to cool a hot car down before you got back to it if it's a hot summer's day. I personally feel that you could probably survive without that ability, as the risks are higher than the convenience. Say a criminal wants to steal your car and they get access to the app on your phone or a cloud-hosted application - they can unlock and start your vehicle without any damage to the vehicle or making anyone think the wiser about them just opening the vehicle and driving off.

They didn't break a window, or slide any tool inside the door to unlock it, they just walk up to the car get in and drive away. Mmmm, that sounds like it was a bit too easy. What about we remove the app and consider the many modern vehicles that are on the road or parked in your garages - they nearly all have keyless entry and keyless start functions. Many car thieves are already taking advantage of this feature with a quick google search I just found the below:

- Keyless Mercedes Stolen without key

The vehicle was stolen from the driveway in under a minute with two devices, a repeater and transmitter. This meant they just went up close to the house and used the repeater to send the signal from the keys inside the owner's home and then was transmitted to the vehicle with the transmitter to get it to unlock, then start before being driven away. It is scarily fast how this crime was carried out. Yes, you could have done a couple things in the above instance to protect yourself - like parking your expensive new vehicle in a garage (I know that may seem ludicrous to some, but could do wonders to prevent this), and store your keys away from the front door like many people do on a hook or in a bowl. Just store them further into the house to help prevent them from picking up the signal to repeat it. Simple but effective methods to reduce your risks of your vehicle being stolen.

Now some of you are probably thinking, what does this have to do with the title and autonomous cars? I was setting the scene for what is to come. Let's look at a Tesla vehicle - one of the most talked about semi-autonomous vehicles. It can basically drive itself in what it calls 'Autopilot mode'. Its vehicles can drive almost completely unassisted by the driver (they say that drivers need to be paying attention and be ready to take back control, but we all know from several incidents that's not always the case).

These types of vehicles are starting to become more common on our roads no matter which country you live in. They will start to get better over time, before we eventually get to the fully autonomous mode in which a driver or steering wheel is no longer even something that is needed. I personally would still like the ability to take back control, but it is only a matter of time until that will not be an option, as the machines become better than us humble humans. They will just pick us up and drop us off. A family will be able to utilise one vehicle that can be summoned after taking someone to work then sent to pick up a child from school or so on, you get the picture.

All of this sounds great right? Fewer vehicles required for a family, meaning less crashes, meaning fewer lives lost. We won't talk about the problems of how a machine should decide on who should die if an accident is inevitable and loss of life is something that will occur no matter the decision – that is definitely a story for another time. Let's just look at the connected requirements a car like this will need, a fast 5G connection (being rolled out in Australia now), or the next generation of wireless connection and what amount of data will be processed in this type of car every minute. The feature I talked about above with a car being summoned remotely to pick you up will certainly make stealing a vehicle a lot easier (just ask it to come to you). Thieves will certainly be able to catch up on their favourite new show on Netflix why they wait for the new car to arrive.

Some cars can already be remotely driven with the apps on our phones, or remotely parked. But what happens when a malicious actor or cybercriminal or - let's get a bit darker and say - terrorists, gain control of these vehicles. They drive them into crowds, or load them full of explosives then drive them into public spaces. That would be horrible, but certainly possible. Hackers and security researchers are already testing and trying to break these systems to ensure that they are as safe as we can make them, but malicious actors will find a way in as no systems are 100% secure (if you think so, then you are truly dreaming).

Remote assassination (drive a car off a bridge, cliff or into a wall at top speed), or abductions, could be completed without even stepping foot into a country in which a target is located. This is a threat that we have no idea about how far it could go, and how to even start to protect ourselves from. But we will need to, or our bright future could get gloomy pretty quickly. I feel we could also have a talk about our connected homes and how that will go in the future, but I will leave that for another chapter.

So yes, as some really gifted hackers on our side are anticipating getting their hands on these exciting new toys, we need to ensure that we as a society know the risks and go into this with open eyes. Not all of these new-found functions will have all positive sides to them.

Chapter 13 - Why is it so hard for us to work together on cybersecurity?

Cybersecurity is a really interesting and challenging industry to work in, but it can be a lonely, isolated job. If you work in a large enterprise or in an MSSP/MSP like the one I do (Davichi), it can be a challenge to get buy-in for security, but this really baffles me (gladly not an issue with Davichi, but still so common). We are bombarded with news chapters after news chapters of breaches that are occurring, I can remember at least three big ones over the last few weeks, but companies do not want to invest or even talk about cybersecurity. Why? I really do not understand this in today's society, everything is connected and online 24/7 and if that is brought down for any reason it can be detrimental to a business's reputation.

Why is security so low on everyone's priority list? Is this our industry's fault? Are the Ghosts of security past to blame? (It's getting close to Christmas and I couldn't help but put that one in) Is security perceived as a hindrance to getting their jobs done, "we just make their lives harder" why should they help us?

I believe it is definitely something that we as an industry have helped to create, we and the ones before us have cloaked our work in mystery and intrigue to I assume secure our place and possibly boost our own ego with being one of the few very secretive cybersecurity professionals or white hat hackers. Yes, security is a tough gig and it takes a particular type of individual to do it right but why all the secrecy? Why not talk about the issues more openly, why not help someone who wants to join the ranks, will it really cause us any issues? No, I don't think it will.

I am still learning (probably always will be) and I always try to help anyone else I can gain entry into our industry but it really seems like I am swimming against the tide. We all talk about the skill shortage and how we need to increase the number of security talent with training or find a way of bringing the talent from unusual sources, but I don't see this occurring in the real world.

Every week I hear of people not being able to get jobs even for the most basic positions in security and some of them are really smart people with some excellent experience within IT or programming backgrounds but just lack the hands-on experience with security. Many of them have done training and are self-teaching which is great but just don't have the on-the-job experience. I know I have said it before, but we really need to look outside of the standard cookie cut certs/experience to get people in the vacant jobs, especially as the gap grows even further.

Now back to the real topic – Why is it so hard for us to work together on cybersecurity? I feel the problem is both internal in an organisation as discussed above but it is also a problem we need to work together to resolve. Chapter one of this book I wrote about managed services being the **next target for cybercriminals** and I discussed that we as an industry need to find a way to better share data on incidents and help each other to be better prepared for security incidents when they happen (Yes that is a when not an if).

Since that chapter was written Davichi and myself personally have tried to push this agenda with several talks with the likes of ACSC and the JCSC (now all part of ASD) and I feel that we received some satisfactory response from them with trying to get this type of open relationship started between industry peers to help protect Australia together (it is always hard to try to get people to see past the fact that some of us are competitors though). ASD has also come out of the shadows during this time and clearly put their hand up to say they are here to help us better protect our businesses not just the big end of town or critical infrastructure (I would like to say we had a hand in getting them to make that move but I really doubt we had much to do with that).

This change in direction and coming out of the shadows is a great move that can only help Australia be stronger and more resilient to cybersecurity threats. With autonomous vehicles and IoT starting to explode we now more than ever need to put aside our differences and start to work together as a country, not just an individual organisation.

I think we can do it, but I just really don't like that it has been more than 6 months and not much has changed. What do we need to do to get people's attention and get them to work together? Are we going down the wrong path here? Should we forget this fantastical idea that we can all come together on this fight and make a difference? I would really like it if we could start an open conversation about this and start to get some real progress or even set up an open session/panel to discuss how this type of thing could work and if the rest of our industry support this initiative.

Tell me what you all think let's put this out on the table and air our opinions so that we can ensure that this time next year I am not piping on about the same problem (I think that would annoy the rest of you as much as it would me) without any progress, if that is the case I will certainly be disappointed but I don't want to give up on this fight as in my opinion it is something that could make a difference.

So, tell me I am dreaming or suggest some ways we could make this work I don't mind either way I just want us all to start talking about it and either concede defeat with this probably over-ambitious initiative or get some progress (I'm really hoping that it is the latter option).

Chapter 14 - Don't let the cyber Grinch get you

With under a month now to Christmas when writing this chapter, many people are starting to attend company Christmas parties and social events. As the date gets closer we all start to wind down for a well-deserved break with our families, so we can come back next year to do it all again. However, there is a lurking threat this time of year that starts to ramp up their efforts as everyone else starts to wind down. Cybercriminals will have started to prepare for an onslaught of phishing (scam emails) campaigns to catch you off your guard, to manipulate your good nature and festive mood.

You will see emails coming in the last days with targeted attacks around last minute payments that need to be processed before everyone leaves on holiday, or the person may be on holiday in which it indicates it's from and they just need you to transfer them $10K, so they can get home for Christmas. Please don't fall for these emails. Check email addresses, names and verify directly with anyone with the contact information you already have (don't use any details in the scam communications as they are more than likely fake). This way if you know the person you will be able to confirm that the information is correct and not a scam that will leave you red-faced when you return to work next year.

You can be certain that you will receive emails around specials that are way too good to be true, yes there are some good sales at this time of year but make sure you do your homework and don't fall for a dodgy email that doesn't match up with a store or online shop you would normally deal with (it may be better to miss a bargain then fall for a scam).

"Two iPhone 8's for $399 – Only for first 20 orders" now that would be a good deal, how about a "HP Gaming laptop at 80% off for today only", I'm sure both options sound very enticing but are unlikely to be real – *Remember if it's too good to be true then it probably is.*

This is just the start. Let me ask you a strange question (Yes, I know you can't actually answer me), Where are all your staff over the festive season? On Christmas morning, what are they doing? Spending time with their families, enjoying themselves not even thinking about what is happening back at the office and that is good - we are meant to be having a break.

Isn't this a perfect time to attack your systems though? A cybercriminal can gain access to company systems, work their way through the systems without any concern of anyone noticing because everyone is off enjoying themselves. Some companies close down for weeks over this time, just imagine what a skilled malicious actor could do with a couple weeks. All your customer details are stolen, they now have access to all of your account information and you probably have saved credentials on the accounts pc (I'm rolling my eyes as I can't believe people still do this) so they login into your bank accounts and start to transfer out money (yes the bank will probably give this back to you but on many occasions the criminal still gets away with the cash), the amounts might be small so they aren't noticed but they add up for the malicious actor.

Once they have all of the data they want and take your money, they will then infect your entire systems with ransomware, this will encrypt all of your company data which they will then try to sell back to you. It's a great business model really when you think about it, they steal your data and make it so you can't use it, then make you pay upwards of $2K-$20K (Sometimes much more) so you can have it back. They cause all the pain but make all of the money while you and your organisation are losing money like a ship takes on water when it is sinking. It's hard to recover from something like that, especially if you aren't prepared.

Do yourself a favour and ensure that before your staff all leave for the festive season (or earlier if possible) make sure that your backups are all working and are isolated from your primary networks (offsite options are great). Make sure you have some good security in place even if that is just a good antivirus (so many people don't even have that – please don't be one of them), if you have time, help your staff by training them on the basics of what to look out for - It could be the thing that saves you from a breach.

Invest the time and money now before the breach takes place, after is too late, trust me. Yes, you may have cyber insurance and that is great and I definitely recommend you getting it if you don't but even with cyber insurance, what about the downtime when you can't service your customers? What about the brand reputation damage? Will customers still deal with you after this kind of incident? We can't try and keep it from them (as I'm sure businesses have done in the past), you must tell them, it's law in Australia.

No systems are 100% secure, we can't stop all attacks but all of Australia's largest breaches this past year were done with known vulnerabilities that could have been patched or secured to prevent them. Yes, sometimes there are reasons that things aren't patched but do the prep work, make sure your systems are secure and you will thank us, that's certain.

Do not leave it until it is too late. Ensure that you have a great festive season, with no hidden surprises upon your return.

Chapter 15 - The Christmas phishing flood is coming

The couple weeks leading up to Christmas is always a chaotic and overwhelming time. You are attending Christmas social events, pulling late night work shifts to try and get your work finished before we all leave for that well-deserved break. Then there is the Christmas shopping, fighting those dreadful crowds to get that perfect gift for someone special. I have to say the lead up is not such a great time (No I am not a Grinch, I really enjoy the Christmas and New Year break with my family – it's just that lead up to the break that really gets under my skin).

The stress during this time of year is not just caused by the above but something even worse, as you all know I am a security professional trying to help our customers keep cyber safe in this lead-up, as this is a time of year in which cybercriminals really turn up the heat.

During these weeks you will all see a flood of system attacks and phishing scams like no other time of year. Maybe the cybercriminals have over-spent on Christmas gifts this year and need to fill up their coffers with cold hard cash again before the New Year, so they can meet their repayments on their mansions or Bentley they just bought and now can't afford the repayments on (Poor cybercriminals). Okay, jokes aside I feel that the reason criminals choose this time of year to launch a big offensive is it is a really good time for them to scam money out of unsuspecting victims.

You are all focused on leaving for your break and just want to be helpful to your fellow staff/customers, so they can all do the same. So, when Alfred or Jo from your Product development team asks you to pay a last-minute invoice, so an order can be completed before leaving for the year you just do it (you don't notice however the email address it is coming from is not an internal address and nothing to do with them at all). What about the accounts team at one of your suppliers reaching out indicating that they have changed the primary account information that they use and to make all future account payments to the new account (you would normally verify the change request with them via phone prior to making a change like this but you just wanted to help them out and get home to get ready for your partners work party – a mistake you will certainly regret in the new year).

Look I know it's a busy time of year and sometimes mistakes are made when we are under pressure, but have you prepared your team and business for the flood of phishing/scam emails they will certainly receive in the next few weeks? Have you done some user awareness training to help teach your team how to pick these scams out of the flood of legitimate emails they are certain to receive?

Yes, I can see some of you rolling your eyes at me or throwing your hands in the air saying that it is too late to do anything now but that is just a cop-out. There is plenty you could still do to help protect you and your team. If you don't have email filtering services to catch the bulk of these scam emails (yes it won't catch them all but better to have it then not have it – Trust me on that). You can still get this implemented if you move quickly and the benefits will be far reaching as you move into the new year.

It's a bit late to organise onsite user awareness training probably (you never know though, it would be worth reaching out to a professional to see if it could happen) but it isn't too late to send around some basic training information to your team that can help them spot a scam email, this is something most providers of user awareness could easily assist you to help your team be just that little bit safer.

Please do yourself and your business a favour and prepare for the phishing flood why you still can, if you don't know where to start reaching out to me or a local security professional most will be more than happy to point you in the right direction. Competition aside we are all here in this industry to achieve the same results, making your businesses safer and reduce the risks of a cyber incident especially over the festive season.

Chapter 16 - Hacked via Fridge: Do you really know what is on your networks?

It's that time of year again when security professionals are preparing for the onslaught of security threats like phishing emails or other social engineering attacks to take advantage of people's good nature or just the fact that most people are busy trying to get everything ready before they leave on their breaks. Many of you will have already read the previous chapter – "Don't let the cyber Grinch get you" (unless you are just picking and choosing which chapters to read on interest, than you may want to go back and read that one) in which I asked everyone to be prepared for the festive season break because it is a very active time for cybercriminals and we all need to be aware of the threats as well as what we can do to better prepare for that well-deserved break.

If we can do it right, we won't have to experience that dreadful call at 3 am when someone reports that your network has been breached and everything is encrypted on Christmas morning. I feel that would have to be the worst present that any CIO or CISO could ever get (if that happens the Grinch certainly found you).

As per the previous chapter make sure you have all the basics covered, have some antivirus protection on your systems (if you don't even have this, then I feel you have probably already been hacked and we should probably have a bit more of a thorough conversation about good security practices). Make sure that you have all of the latest patches, staff know what to look for with regards to scams (or at least know if it looks funny that they should go to your team to look at it – feel like they can approach without any condescending attitude).

If you have all of the basics covered that is great we can create a good security posture from that baseline, but this is just a starting point and we should never stop trying to improve on our security as malicious actors/cybercriminals won't stop trying to break into it. So, doesn't that make sense that if they are not going to stop trying we shouldn't ever? It does to me.

So, with that in mind do you even know what is on your network? Have you done in-depth scans of your networks and laid a hand on every device? Can you tell me that every device is controlled by your team and is as secure as it can be? Most of you will have not answered yes to many of these questions and that is fairly normal. Most of you will have BYO mobile devices, laptops that you allow staff to use on your company WIFI which is a calculated risk on its own that your organisation needs to manage with mandatory protections (AV provided by business or can't connect to the corporate network is one option companies use), Guests/ client devices or similar depending on your organisation (motels will provide Wi-Fi to guests – I really hope they are an isolated network but I bet there are a few of you out there that just let them all on the same corporate network).

They are the most common but with the advancement of IoT (Internet of Things) do you have an internet connected fish tank in your company foyer, what about a smart fridge in the company lunchroom. Did you know that you could even get a connected fish tank, the thermostat can notify the owner of the temperature is abnormal, the filters can signal if there is a problem that needs to be rectified and a malicious actor can use it to run a script on the network or do some investigations without any security sensors being tripped, Do you monitor your fish tank filter? Should this even be connected to the corporate network, no it shouldn't? Maybe the guest network that is isolated.

What about that smart fridge, that someone thought was a great idea to enable the milk supplies to be automatically ordered from the local supermarket? Firstly if I was just wanting to have some fun I would order 50 bottles of milk and or something really weird that would just irritate the owner of the fridge (50 punnets of strawberries and 20 packets of dipping chocolate – I am sure that would raise some eyebrows in accounts when that bill came through) I was able to take control of but if a malicious actor got in and could stay undetected for months undetected gathering information on the networks weaknesses before launching their offensive, by the time you realised it was happening it would be way too late.

Just think about it, you spend 1000's or maybe even millions on securing your network but the $200 connected device with no interest in security during development and no way to update them (why would they add this feature these are just supposed to be mass produced and cheap to deploy. They don't care about if they are secure that's not what they are trying to achieve but we need to change that opinion long-term or when every device has a connection on the internet we will lose control of our own systems (yes, we probably already are starting to lose control).

I can see you rolling your eyes, no one could breach my network via a fish tank... think again, do a search for "Casino breached via fish tank" you will find many chapters on a casino breach that occurred via the heater in the fish tank in the casino lobby. No one would have ever thought it was even possible and to be honest their security team probably didn't even know that it was on the network.

So, let me ask you all again, do you really know what is on your networks? If not, find out and secure them as best as possible maybe even disable the connected features that you really don't need. Do some full scans of your networks and find every single device, know what it is, where its located and why it needs to be connected (if it doesn't – you know what to do). Once you know everything you have on the systems you can better make decisions on what risk you feel is acceptable and create plans that will ensure your systems are more secure than was ever possible before.

Knowledge is key if you don't know what you need to protect how can you truly believe your systems will ever be secure. Do the basics, know your systems, make sure you have at minimum basic protections in place and update your systems. Then reduce what you allow on your network if it isn't used anymore or doesn't truly need to be connected to the outside world or have access from external sources take them off the network entirely.

Chapter 17 - Would you know if you had been hacked?

Let's say I was a cybercriminal or Blackhat hacker (The idea of being a black hat hacker is humorous to me, it would be a stretch for me to even be a Grey hat but let's go with the story) and I am starting my week with considering my latest target to loot. After a bit of a search on google, I decided to attack a mid-sized enterprise in Australia. I look up the company sites and do a quick recognisance on what they have as a publicly facing facility. They have the usual with websites, customer portal login, contractor's portal login, remote access server for staff, email servers and enough ports open to internal systems that would give several options of attack.

I decide that I want to investigate further and that I would look up all the company staff on social media to determine who would be good targets and collect further information that could be useful in determining a user's account or password. I could start with pet's names, children or loved ones and which targets would give me good systems access if I were to guess the correct combination. Some people do really make it easy for us to get in. Now please listen to me, if only once listen to me now. DO NOT use a pet's name, family member or loved one's name combined with yours or their birthdate. This is the easiest information to find on Facebook or LinkedIn.

Before its time to move on to the actual attack stage, I look at all the recent ICT related job advertisements you have posted. This tells me what server platform you use, the firewall, application suite and any protections in many cases as you ask people to apply if they have those skills. So, I know what vulnerabilities I should narrow it down too during my attack (thanks for saving me days trying to footprint your systems).

Now there would be a few more steps like scrapping DNS and so to ensure I have as much information I want before doing a deeper look at your systems but let's say that I am ready to take things to the next level. You and your organisation have absolutely no Idea I have been researching your systems and if I am being honest there is no real way as I (the malicious actor/Cybercriminal) haven't really touched your systems, no active scans that would alert you that someone was poking around. Just normal everyday traffic that would look nothing unusual. However, I know who the primary targets are, email addresses, family members, pets, birthdates and if I have made some real effort in my information gathering I will know what the staff general habits are. When you are online, what information you normally share and who my weak link in the chain is – **My primary targets.**

At this point I have two main paths I could take, the first would be to attack your systems the good old fashioned way, scan your systems find holes in your armour or gather information together to generating passwords that you may use. Which in many cases you could just guess based on what can be found with your Facebook page. People truly do share too much on social media and an attack can be based entirely from information that has been found here, it could even help me find when a good time to attack is (you have just posted that you are on holidays for three weeks so you won't notice that I am poking around in your things. It's unlikely that the IT team would notice either as many IT departments are stretched so thin that they don't know who is out of the office and shouldn't be working – we can't really blame them it's just how things are these days).

This direct method would be a risky option and could get me caught before I even get started so option two would be a much better option. Social engineering is one of the most used methods to gain access to desired systems, I could go down the phishing path and have you click on the link to update your password or login to fix your storage space issue on your mailbox or your user profile. I could even ask you to login to your Apple account or google account to verify your info as a suspicious login has taken place (Not yet anyway, that will happen after you give me access without even realising). In this instance, you will see all the normal screens and won't even know I have done anything.

How about I call you saying I am from your IT support helpdesk or external provider and I need you to reset the password now (then tell me what it is, so I can test it ?) Thanks for holding the door open for me to walk right in. So, I am now into your systems from one of the options and I can move around the systems why you are on leave. I would then gather any information that would be useful for me or that I could sell on the dark web to someone who could have a bit more fun with your details (maybe even use the current passwords to login to your other accounts as you probably use the same one on everything – I really hope you don't but I am sure there would be a least one of you).

I would now cover my tracks so that you would not even know I had been in your systems (many companies never know they have been hacked until it's too late). How would you know that I had been there, no logs to indicate I had, no missing information, not even a whisper that something was wrong? The only way you know that you have suffered a breach is if information that is used in another attack or financial fraud is linked back to your company or a data dump is found for sale by authorities on the dark web or worse you arrive at the office one day and the entire systems are encrypted by a ransomware virus.

I would do this if I thought this method was my best bet to make money from your organisation, many organisations do pay and the price is great for me in the range of $2k-200k depending on your business and how valuable I think you will class your information and the consequences if you can't get it back. If you pay I would be super helpful if not you would never get the information back.

So, what Is my point? Practice safe internet usage, don't use the same passwords in multiple places, don't use your pets or loved ones as part of your passwords. Use a catch phrase that is long, forget all the letter, number, symbol combination that we have preached for years. A long simple passphrase will take so long to crack that it would not be worth my effort to get in and when I finally did it wouldn't matter anyway as you have probably changed jobs or retired (that's how long it would take me to crack it).

Don't put all of your information on social media, some things should always be confidential so use some self-control and not overshare. Minimise the specifics you put on job adverts (don't make my job as a criminal to easy) and honestly just consider emails or phone calls when you receive them don't just trust that something is legitimate as in many cases these days they are more likely a scam than real.

Okay so final note – I am not really a malicious cybercriminal I am one of the good guys (or at least I try to be) out there trying to help you protect your systems from the real bad guys and girls who definitely don't have your best interest at heart. Do yourself a favour and practice good security, it will be a better result for all of us except that cybercriminal, no new Ferrari for them this month.

Chapter 18 - Cyber warfare is here and we are not prepared

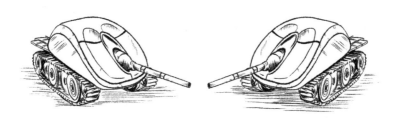

Over the last few years there has been a massive increase in the number of breaches occurring globally, every time we look at the news another high profile company has been breached and customer data leaked all over the internet or for sale on the dark web. This is sadly our new normal, which is a little concerning but nevertheless true. However, I believe there is a misunderstanding by many that all of these high-profile breaches are committed by cyber criminals, many of these attacks are actually carried out by specialist state sponsored hacking groups. It is their sole responsibility to attack foreign countries to collect information that will benefit their own government.

War is no longer waged with troops but with the swift execution of a cyber-attack that could bring down a whole country in just hours without even setting off any bombs, firing any guns or stepping foot in your enemy's country. Yes, there will probably still be soldiers invading a country once the initial attacks are completed or maybe they will just leave the country to tear itself apart. Seems like a safe way for a country to take out an enemy doesn't it?

If a country such as North Korea wanted to take out say China (Which they wouldn't because that is where they get most of their supplies from – but you will see where I am going with this in a minute) being that they are a smaller country and would not have as much military firepower it would be a massive mistake for North Korea to attack China in the traditional sense with soldiers and missiles (you get it) they would certainly sustain many losses.

North Korea, however, has a state-sponsored hacking group generally known as APT38, this group is believed to be used for both building up the country's financial reserves (for which they are pretty successful at doing) and for attacking the enemies of their leader and country. If North Korea truly wanted to attack China and truly have a chance to win, they would need to use the strongest weapon they have that could inflict the most damage in the fastest possible period. Attack with all the force and abilities of APT38, take down the banking systems, shut down the power grids, and interrupt all of their communication networks. Kill their way of life, no money, no food, and no communications. It would not take long before society would start to implode and people starting to riot or pillage to survive (it would get ugly pretty quickly I think).

NK would just have to sit back and wait for this to happen and when everything was on the brink of complete collapse, come in with all their military strength and bring order back to the chaos. War one with minimal cost and damage to their own forces. Seems like a smart option for me. I know what you are thinking this couldn't happen China would be too strong and they would have all of these systems well protected. Yes, they would have all these systems well protected and they probably have a strong cyber defence capability but have you seen the news anytime over the last year? Many companies and government organisations all over the world have what they think is a well-protected system but every system has a vulnerability you just need to find it.

Therefore, they get their hacker army to find these weaknesses, plan a simultaneous attack and initiate them all at once. What systems they did not bring down would probably collapse under the load if they were able to take out enough of them. This sounds very plausible to me, and in some form, we are already seeing attacks of cyber warfare with one just before Christmas (2018) in which China (Allegedly) targeted International MSP's in order to gain access to their customer's systems and data. It was a successful attack, which until then had not really seen in such a large scale but this proves my point that a large scale synchronised attack could be very successful.

What about as the IoT and smart everything invades every aspect of our lives, they could start their attack via infecting all of these devices in which have little if any security functionality and spread like wildfire through our entire lives. Yes, governments have probably been thinking about this for years and I am sure they have capabilities in place that would amaze me (Or at least I hope they do) but this type of attack will most likely target the private sector not the government or military systems (at least not initially) but are we prepared to defend our systems form such a disciplined and well-skilled enemy?

I am not sure we are. We can barely protect ourselves from cyber-criminal groups who just want to steal our money so how can we say we are ready to defend ourselves from true cyber warfare if it came at us with full force. I think we would lose in a spectacular ball of fire (just to be dramatic) and it is scary to think about as if we took down all of our electronic systems many wouldn't know how to survive, how to keep our families safe and warm or even feed them.

We need to work together more, know what is on our networks and at least cover all of the basics so that we can be as protected as possible. That way we might just have a chance to defend ourselves from such an attack as sadly I think it is truly just a matter of time before it actually occurs, and I hope that Australia is not its target, as I know we are not ready for it.

Chapter 19 - Post Grad qualifications: Are they worth it in Cybersecurity?

It was mid-2013 when I found myself looking over options for post-grad degrees in IT, I had been in IT since I graduated high school (1999 – wow that makes me feel old) when I started a basic technician job at a small IT services company. I stayed in that position for five years and made a move to the city to expand on career opportunities. However, I found that it was a little difficult to get a good job because I didn't have a degree, I was confused and couldn't understand why I couldn't get a job just because I didn't have a piece of paper to say, that I could do what it is I could do, in many cases much better than most candidates at the time who had actually completed a degree. It was just perplexing to me.

I did do well and scored a pretty decent job with Gold Coast Tourism as an IT Support Engineer (It was an awesome place to work, People and culture were second to none), but it has always bothered me about that piece of missing paper. I moved on and after a few years was invited to come back and run the IT services company I had started out in all those years earlier. It felt pretty good to have earnt that respect from my old boss and be given such a great opportunity to manage and provide my input to an organisation which I had such fond memories. During my time as the General Manager, I had the privilege to be able to help some great young techs start their journey in the industry and after a while, I came back to that thought train of my missing formal qualifications and decided that I wanted to fix it.

I started looking at the options available for bachelors in computer science or management and they all seemed to cover knowledge I already had from actually doing the work, instead of just reading about it. So a bachelor was out for me, it just wouldn't be worth the money and effort in my opinion but then I found some information on a couple different masters programs that would consider my experience as part of the entry requirements and would allow me to bypass the undergraduate study and enter via a graduate certificate arrangement (If I successfully complete the 4 units of the grad cert they would allow me entry to the masters degree with full credit of the four units towards it). There were a few places that offered similar programs, but I personally chose the IT Masters program through CSU (Charles Sturt University).

IT Masters and the CSU's partnership in my belief were a little in front of the curve with their offering, as it was a mix of both industry modules with academic modules which would help me improve academically and expand skills/abilities (really liked the industry units). I chose to go down the Master of IT management degree path as I was working as GM at that time, but I had been interested in IT security for quite a long time so just for fun I chose digital forensics as my major.

That choice of major was a changing point in my career and it is what gave me the security bug once and for all (I was well and truly hooked after my first security module). I graduated with my first masters in 2016 (yes, my first – I will get to that in a minute) and I felt that the foundational skills that I had learnt coupled with the industry related skills were worth every bit of my time and the cost (I still think university is ridiculously expensive but we won't get into that argument now). It really did improve me as a person and a manager/IT specialist. I finally understood why people wanted that piece of paper.

In 2017 I decided that I wanted to dive deeper into that security rabbit hole, I had caught the bug for back in 2013 and signed myself back up for the master of information systems security (Again with CSU and IT Masters – I thought why change it If it worked well so far). It's been a little tougher this time around with the arrival of my first child but possibly more rewarding because of the extra effort required to get through it each session (certainly not a cake walk that is for sure with full-time work, expanding family and a pretty demanding study schedule). Please take my advice and don't try to do two units at once it is a pretty horrible experience (I think my wife thought I was a zombie some days during that regrettable session).

Now I understand that I have gone off topic a little, giving you a bit of my history but I wanted to set the tone for my personal experience which I'm sure many of you in the industry will find similar to your own.

I have been in dual security and IT roles for 7-8 years now with a move to a primarily security position almost two years ago now, I honestly believe that both the first masters and my current one made me a much better leader and security professional. Not just for the security skills that they helped me to develop but the improved writing and communication skills, helping me to think outside of my own world and the ability to go out and find answers to the problems I faced is undeniably a huge asset to me personally but in my career moving forward as well.

Successfully gaining my current position at Davichi was certainly helped by these qualifications, coupled with my hands on abilities (mostly self-taught while doing the degrees) in both security operations and penetration testing. Without these qualifications, I would not be in the position that I am today that I can say for certain but I want to make It clear this is about more than just a piece of paper it gives. The security bug, of course, gave me drive to push my boundaries and learn more of my current skills but the education has changed me for the better and I feel that they are definitely worth the pain you have to endure to gain them but is that enough?

Some in my industry won't agree and say that industry certs are more important than a university degree. I disagree, I feel both play a big part in our industry and it would be my recommendation that anyone looking to gain entry should not focus on just one but actually gain both through a program like I have chosen or separately if that works best for you to ensure that you get a coverage of both the formal and industry skills. Do not narrow your vision, see what is out there and do what feels best for you.

In my case, I feel that signing up to the degrees via CSU/IT Masters was the best thing I have ever done but what you need to figure out is where you want to go with your career and will the skills you get from either post-grad degree or an industry cert make you more desirable to hiring companies or will it do nothing to get you into the position you want to ultimately end up in? If it isn't going to get you where you ultimately want to go, then you might as well through your money out of the window of your car moving down the highway, it won't benefit you or anyone else on that highway (might cause a few accidents if you throw to much out but certainly no benefit).

Chapter 20 - Mental health: Is it a big issue in cybersecurity?

This chapter is going to be a little different to my normal style. I really want to do my part in highlighting a problem in the security industry that we all need to consider and discuss openly so we can help each other in tough situations. Stick with me and let us advance this conversation in our circles, it will be a benefit to us all.

I am a security professional / would-be ethical hacker and I feel that the glorified image our society has of my profession is all wrong. Movies and TV shows have depicted us as nerdy types who have little to no social skills gamers (I am personally not much of a gamer) and general techies who wage battle with all sorts of foes to fight injustice in our world. One day we are taking down big evil conglomerates or some mythical bad people that save us all from the impending doom.

What about Abby from CSI who can bypass encryption on suspects systems and have all the answers they need to crack open a case in minutes (it is not that easy, we wish it were). We then would have dramatic battles in cyberspace to win the fight (if many of you saw how it was really done you would lose interest quickly but Hollywood does this very well). Okay, I admit I am being a little dramatic here in trying to paint a picture of this mythical creature they call a cybersecurity professional (Going all Hollywood on you).

What we really do most of the time is dredge through millions of logs (if we are lucky we have a SIEM which makes that much easier), respond to the flood of alerts from port scans through to a stopped email attachments that may be malicious. Our job is mundane most of the time and really is not so Hollywood dramatic. We have a workload that never ends and in many organisations, it actually seems to be climbing constantly which makes you feel like you are fighting a losing battle. Yes on occasions, we get to do penetration test engagements or red team projects (where we get to pretend to be the bad people) and get to have a bit more fun.

On those rare occasions, being in a security professional or ethical hacker is awesome but most of the time we can be very isolated and have minimal interactions with the rest of the world. We can work 60+ hour weeks and just honestly are pretty stressed out with the avalanche we call our workload. These issues are exasperated by the skills shortage and the seemingly difficult to clear roadblock for people to join our ranks. I am not going to talk about that issue in this chapter as I have already covered that with my previous chapters "you want a career in Cybersecurity, are you crazy?" or "What to look for when hiring Security Talent: Hidden talents".

So what does that give us? I feel it gives us an industry full of individuals who have a high risk of mental health problems. High stress and minimal downtime certainly cannot be good for anyone, even for a short time frame, so it should not surprise us to know that cybersecurity professionals would have a high burnout rate, should it? A CISO is said to only have a 2-year shelf life and would normally burn out after that time, this is a big problem and one that needs to be addressed.

How can we handle this problem? To be honest I do not know but I think we need to really try to help each other out more and help ourselves be a little less stressed. If we see someone that looks to be struggling, ask him or her if they are okay. Just talking to them may be enough to help someone through a tough time, it may not be security related but that does not matter.

I am not a mental health professional and I do not know how best to help someone in this type of situation but I can listen and suggest they should talk to someone who is a professional, it could save their life. Mental health is not something that we should be embarrassed by and push under the rug so to speak. Let's bring it out in the open and work on this together.

I wanted to write this chapter to help bring this topic in front of as many people as I could and I was inspired by Simon Harvey who is a strong advocate in this particular space and fellow security professionals. I have been lucky enough to see Simon speak on the issue on a few different occasions and even luckier still to have been at our Brisbane AISA branch meeting for his first presentation on the topic which seems like such a long time ago now. He openly talks about his own personal demons from his past and I feel that he should be given kudos for putting himself out there like that to try and help us all be there for each other. I hope you do not mind the mention Simon and keep up the good work.

We need to try to expand the intake of new security professionals and learn to take some downtime for our own sakes. Skills shortage or not there will not be any of us left to defend the networks if we do not do something. I am in a good spot mentally and have a good balance of both work/family life but not everyone does, so let us do our part to make life just that little bit better for us all.

If you need a hand or just want to talk, reach out to someone for help you may be surprised how much people are willing to help, we are not alone in this world, so you do not need to feel like you are.

Chapter 21 - A Hacker, I am

This had been a long gig, with many long nights just crawling through logs, or reconnaissance data but as I sipped from my cold cup of coffee, it clicked. I had access. Finally, all my effort paid off. I spent what seemed to be endless days trying to worm my way into my targets systems. There were days I wanted to just bang my head on the table from the frustration of it, I needed to get in, and failure was not an option.

I started by foot-printing my target so that I knew as much as I could about them. I even set up social media accounts and scrapped all the information I was able to generate a list of accounts and passwords that might get me on to their network. I needed to know everything about anyone who worked for the company, how many children or pets they had, their names and ages. What they like to do, what they like, even what they eat.

With my new-found knowledge I poked and prodded the organisation's protections trying to find my way in, that crack in the armour that would be their downfall, but I couldn't find any. I had to be careful not to do something that would alert them of my approach, soft touches so the firewall or IDS wouldn't see me as a threat and alert the IT team that the defences were being tested. If they caught me it was all over.

Every touch I made left a trace. Yes, I cloaked my systems with several layers of anonymisation, but one mistake could be enough to get me caught. I was getting closer, I could feel it. Just another couple of hours and I would have the access I needed to own my targets systems. They would be mine.

In the end, it was social engineering that got me access. I had sent phishing emails to all the staff to try and get one of them to click on my link and allow me onto their system. I had to refine my efforts and sent four different versions to different targets, Specific employees who I thought would be my way in for one reason or another, sometimes just a hunch on someone that I get from their social is enough to put a target on their backs.

The victim that got me in, was chosen because they would give me the access I needed and allow me to work unabated. Their social media had told me they were leaving in two days and would be out of the office for two weeks why they were in Bali, soaking up the sun. Perfect target for me to work my magic. I sent them a nicely crafted email from the IT department asking them to reset their password due to suspicious activity on their account via my conveniently added link that would take them to reset page, once they gave me the details it redirected them to a "failed to change" page and asked them to press ctrl-alt-delete on their systems and click the reset password option then re-enter the new password info to ensure that it takes effect.

Password changed, they gave me a copy before they actually changed it and they think everything is now all safe (they were wrong). Now I could have gone crazy and accessed the systems right then and there but that would have given me away. I waited until they went on leave (they told me for certain via their social pages – People share way too much on there).

I was in the network and no one was any wiser, I had valid credentials and there were no red flags being raised. IT probably didn't even know the user was on leave, so why would they even look at the account access twice. I dug through the systems and took notes on what I found for future reference. I had access to everything, once past the outer wall protections nothing had been implemented to restrict my access. A mistake on their behalf.

Now it was time for the boring part, writing my report on what had occurred and outlining the weaknesses that had led to me gaining access to their systems. Yes, you have probably now guessed it I am a Pentester. I had been paid to break into the organisation's systems in secret, to break what is not supposed to be able to be broken. To then help them fix what I was able to break in the first place. In some people's eyes that would make me a hacker. Which I guess in a way it does but in today's society the word hacker has been made out to be a hooded figure hunched over a keyboard with some weird matrix code running down the screen while they steal all your secrets or drain money from your bank accounts...

That is not what a hacker is, hackers are normal people who in most cases are the ones who are actually defending your networks, not cybercriminals who are the ones that have been wrongly defined as hackers for too many years now. Now to be clear, even the malicious actors and cybercriminals or even the state-sponsored hackers don't hang around in dark rooms with dark hoodies on while hacking your systems (okay, so there is a slim chance that one or two might do this but very unlikely).

Walk through any public place and you will probably see a hacker. Maybe on your way to work on the bus or train, in the car that was stopped at the lights next to you when you turned into your suburb or dropping off your kids at school. We are everywhere. Now let's clear something up, I am a hacker and I barely ever wear a dark hoodie or hang out in dark rooms unless I am watching the latest movie at a cinema or sleeping (something we all need to do occasionally).

So how about we all start thinking a little different about who or what makes a hacker. Secondly let's start calling cyber criminals exactly that cybercriminals, not hackers. Yes, I know it's not as dramatic or sexy as calling them all hackers and yes, I know the media has already made everyone believe that stereotype hacker in a hoodie scenario but let's do one thing to make our society better and tell the world we are hackers and remove the stigma that comes with the name.

Apologies for going all Hollywood at the start of this chapter but I thought it would be the best way to draw readers in and maybe drive home the message, **I AM A HACKER** (not necessarily a very good one but I am) and that's okay, I am one of the good guys. I would love it if the worlds security folk stood up and said it for all to see that we are hackers and we are here to help, not all of us are bad guys you need to fear. You have probably even met some of us and didn't think twice about it. We may even be your neighbour or a part of your family. We are not the bad guys.

Hacker is just a name, don't judge too fast.

Chapter 22 - A day in the life of a cyber-criminal

I lay awake in bed looking over at the alarm clock on my side table it indicates 5:59 am, I am waiting for it to click over to 6 am and the alarm to sound. **Click**. The alarm sounds and I spring into action and the alarm was silenced as fast as it started. My wife is still asleep and barely stirred from the sound. I walk down the hallway past my children's rooms and head for a shower.

Now I am all dressed and prepared for the day, it is time to make a coffee for my travel mug and head off on the commute to the office. It is a long commute to the office today, there must have been some sort of accident as traffic was crawling for such a long time I was almost late to the office. I am lucky it's Friday, I am looking forward to getting away for the weekend with the family, and it has been something we have been saving for, for months now. A quick escape down the coast will do us all good. The pressure at work is starting to get to me; I just can't seem to meet my targets but today just might be my day.

I see Harry and Jane talking at the photocopier as I enter and say a quick hello before making my way to my cubical down the back. It is a small space and looks just like the other 50 cubicles on this floor except I am at the end of the row and have a direct view out a large window just next to my desk which overlooks a man-made lake that is quite nice to lose myself in for a few minutes every now and then during the day when I am just not having a successful one.

I power up my computer and get ready to load up the contacts database I will be targeting today, I have a new group which I hope is much more fruitful than the last as my supervisor is just riding me to get better results and there is no way I will get the end of month bonus if I don't pick up my game.

I load up my email template tool and decide to go with one of our new templates; I just have a good feeling about this one. It is a "you need to reset your password due to suspicious activity on your account" template, these are normally quite effective and it will take the recipient to a page they will enter the old password and then get them to add in a new one, it will then fail and ask them to press ctrl-alt-del and select reset password, then enter the detail in again just to ensure that the change takes effect. I send this out to 50K recipients on my list and I start to get some response almost immediately. I start to correlate the details into a working sheet that will later be injected into my tool to spread ransomware onto the target networks.

I get a few thousand responses and import them all, so that I can inject the ransomware executable onto all of the systems. I hit the run button on my campaign and it starts to do its thing spreading my bug through their systems, but this will be a bit of a wait now. I will go down to the café on the ground floor now for some lunch and will see what the result is on my return. I don't want to get too excited, but it is looking like my luck may have turned today but I will be patient and just see how I go.

Lunch was nice; I caught up with some of my old colleagues from the information gathering team that I worked with prior to moving to the attack group. It was good to just have a quick catch up and see how their families were all going. I get back to my desk and I have hit the jackpot. I have 100 organisations that have been infected with my ransomware bug and some of them are really big fish that will definitely get me some money though, I can almost smell the end of month bonus.

I look through my encrypted message app that all the ransomware decryption requests will be sent through too and I have some sitting there waiting for me to respond with how much it will cost to unlock all of their files. I request more information from them with how many machines that are encrypted and I have a hospital that has 300 machines locked and another accountants firm with 80 systems.

I will demand 10K bitcoins from the hospital and 5K from the accountant firm. I get an almost instant response to my price with the hospital asking for a discount, as they cannot afford that price, I send back 8K is my lowest offer. They accept and I will now wait for their payment before I send through the decryption tool. The accounting firm did not even negotiate they just paid the 5K. A few minutes later, the hospital sends through payment as well.

I send through the keys and the decryption tool so they can get back the files and offer help if needed to restore them, but they do not ask for anything further. It is now 4 pm and I can see that I have another four requests for decryption prices come through for some smaller under 50 user sites, so I send them through quotes for 1.5K bitcoins for decryption keys.

It's time I start to pack up for the day and make my commute home, it has been a successful day today and I will surely be getting a few thousand bonus this month now at a minimum. My wife will be happy, we have school fees coming up for the kids. A good day overall. Time for the break with the family...

So I bet you are all going what is this guy talking about, why is he telling us this story of what this criminal is doing, it's because I wanted to paint the picture for you that these ransomware syndicates could quite possibly be run like a standard 9-5 job in which the workers get up and go through their day just like many of us would. To them it may be just their normal job and they do not see it as a problem. They get paid a salary with some bonuses if they can scam us out of more money than expected. They have families that may not even know exactly what they do (they probably think they have an IT support job).

With the way they work, this scenario could be quite real and not just a fantastical tale I have spun. How can we stop this type of criminal organisation? I don't have that answer and it will be a combination of security folk like us finding them and law enforcement going through the right channels to get them to shut down. In this kind of scenario, I'm sure their families would see us as the bad guys not them, this is how they survive, feed their families.

Other employment sources need to be created in these places to help steer people to legitimate careers and stop them from becoming part of this horrible scenario. Better for them and for the rest of the world don't you think? Well, that's enough of my story for now and I hope you liked the slightly different style I took this time around...

Chapter 23 - Physical Security being overlooked

The next time you are visiting any business, private or public look around and take things in with a – let's call it "security filter". Really look at your surroundings, every detail and I will bet you will see some obvious things that would have a negative effect on IT security.

Let's put some more context to this so you can get a better picture of what I am talking about, you walk into a shopping centre and pick a retail store (Any store it doesn't really matter which) and look at the counter. Many places will have a PC sitting on the counter, sometimes nice neat AIO's and you may see a monitor sitting on top of a slimline box or even worse a mini pc mounted to the back of the monitor. Okay, so most of the time cables are well controlled and it all looks great but what about the security of this device?

I was in a shop last week (Let us call it an auto parts store) and I asked for something and the gentleman who was assisting me went out the back to see if they had what I wanted (this is a common scenario). I was out the front of the shop all alone; I could not see any cameras or general surveillance. Even if there was I could have easily leaned forward, resting on the counter and plugged in a compact USB stick into the back of the computer, they can be smaller than a 5-cent piece nowadays. With something like that, I could have executed an installer, recorded keystrokes (waiting for me to collect or send them to me by the pc internet connection), there could be some interesting options.

This is not something that people think about; it has probably never even been a thought at all. What about when you go see your accountant and you can see a post-it note stuck on the monitor with Password clearly written for all to see. (I have actually seen this one and had a conversation with an accountant about this exact scenario, who once I explained the risk, moved it into a locked drawer in their desk – an improvement I guess *rolls eyes*).

A few months ago, I had a sick family member and had to take a trip to the hospital (they are all better now in case you were wondering). The same thing, PC on the desk USB ports available and worse a communication cabinet in the triage/treatment room with a key still in the lock with network switches and possibly a router of some sort there for me to access. Again, no camera in the room - that would be unethical given that it was a treatment room. We were left alone for more than 30 minutes at a time in the room (Do not get me started on the ridiculous waiting times at hospitals – 4 hours before we were even seen). I could have easily accessed the network and who knows what sensitive information I could have gleaned from the network by just listening to traffic before even considering what systems I could break into.

Yes, in some cases, USB ports are disabled (I would almost bet that none of the above is but hope the Hospital has this implemented) and this would prevent me from executing something from a USB port. I could use a keylogger that plugs into the port in which the keyboard is and then the keyboard would be plugged into it just like a USB extension (obviously shorter and much more malicious). The machine wouldn't see it as any different and I would be recording all keys that were typed.

Let us look at a different angle now, do you have a waiting room or reception area at your business with seating that customers or guests to your business wait for a meeting or the next available representative of your business? Do you have network ports available in this area, if so are these ports live?

Could I just plug my laptop into this port and have access to your corporate network? If this is the case at your office ensure that these ports are unpatched at the switch, so this can't happen, best practice would be to only patch ports in when they are needed not just make all ports in the building live all the time. Yes, I know that it's a pain to get IT to patch it back in or annoying to have to do this when needed but it is definitely better security practice.

I think you get what I am trying to get across to you all by now (or at least I hope you do). Take a few minutes and just look at your location with security in mind. Can you see the obvious things like those that I have pointed out above that could put your business at risk? Really, think about this and fix the problems, move the pc under a counter, disable ports on the machine in the bios so I cannot just plug something in. Unpatch the unused ports in meeting/waiting rooms, take the keys out of the server and communication cabinets (This is a stupid thing to do, so go take the key out of the lock now and secure it somewhere a little better please).

If you do this, you could be dramatically improving the security of your system and together we would have done at least one thing that could make a difference (That is something to celebrate). One small change could prevent the embarrassment of needing to announce to your customers and possibly the public that your systems were breached and all of the sensitive data you have stored was extracted from your systems. This type of incident could be the end of your organisation, no I am not being dramatic it really could have that great of an impact. If customers do not trust that you will keep their data safe, why would they do business with you? No customers, no business.

Chapter 24 - Ransomware, the recurring revenue model for cybercriminals

Ransomware will make even the most experienced IT professional shiver. Let's say you woke up feeling refreshed and have a perfect commute into work, Life is good then when you walk through the office door one of your team walks up to you and tells you that one of your sites has been hit with ransomware and everything is encrypted. (that's a Holy Crap Moment – Day ruined).

Hopefully, in this situation, you have isolated segments of your networks and backups are all UpToDate (and on one of those isolated segments) so that you can recover with minimal data loss. Here's a scary statistic about backups – only one-third of SME's indicate that they continuously backup their systems. That is bad but let's add in some statistics about cybersecurity incidents.

One in ten SMB's have been affected by ransomware and 516,380 SMB's fell victim to a cyber incident in 2017. This figure is sure to be rising over the last 12 months, as the notification of breaches is constant, and it is almost impossible to go through a week without hearing about one form of a breach or another. Let's add in the average amount that the majority of SMEs would need to pay in ransom to unlock their encrypted data, $4677.

Even if only 25% (129,095) of those victims paid the ransom it would equate to $603,777,315. That kind of figure would certainly fund a couple of mansions on the beach in different sunny islands around the world. What about a Ferrari as well at each location in a couple of different colours so you can choose your ride depending on your mood? Sounds awesome, doesn't it? I am sure that there are plenty of non-extradition countries that could meet your needs.

The proceeds are an estimate and could be much higher, however, it would be split between thousands of cybercriminals, so the mansions and Ferraris may have to wait but even $300K-$500K annual income would be a decent salary. Invest that back into some legitimate income sources and you could, in the end, afford the desired fancy cars or estates (that is if you are not caught or accidentally travel to a country that has an extradition treaty – then you may lose it all).

Why is Ransomware so popular? Simply put it is the cybercriminals recurring revenue model. Look at it in the light of office 365 for Microsoft, Microsoft will sell you 200 licences for office 365 and provide you access to the email hosting services for $17 + a month per user. They already own the platforms, it doesn't cost them anything further to provide it to you and they continually earn money from it. Smaller payday initially but long term they will make a killing from it.

If we go back to look at ransomware, a malicious actor will develop a ransomware virus which could take a few months to get right if they are a skilled operator or possibly they may just buy it from another party at a fee (price will depend on how good it is and if it has been seen before). Once they release the ransomware bug they will just start to receive payments from their victims and all they will need to do is provide unlock keys and move around their money, so it is harder to track.

After the initial costs, they will get a constant revenue for very little effort, let's say the bug stays effective for a 2-year period that is a good return on investment no matter how you look at it. It may even continue to earn income for years like some of the more successful variants to date.

I hear peers say that ransomware will soon be a thing of the past, but I really don't think it will, why would it be? If this recurring revenue model that is ransomware can continue to generate money, why would malicious actors stop using it? Even if they diversify and do some good old-fashioned hacking and sell stolen data or credentials on the black market or build a network of crypto minor zombie machines it just makes sense that if it still brings in more than it costs to run then they will continue to ride the gravy train we call ransomware. So, in my opinion, it will still give me that shuddering feeling for many more years to come, until they develop something else that creeps in the dark and keeps us awake at night.

That's a terrifying thought - what could be worse than Ransomware?

Chapter 25 - IoT - Your security Nightmare

IoT is the latest buzz topic that is spreading like wildfire, almost everywhere you look someone else is discussing IoT and what it is going to mean for us as a society, the benefits that it can bring but what about the negatives? I do not want to be the bearer of bad news but IoT is going to be your biggest nightmare. I bet many of you are sitting there reading this and you are thinking how is IoT a problem? Well, let us look at IoT in our society today and then what it may look like in a few years time.

What are the IoT devices that are out there now? They are starting to pop up in places you would not believe with smart cities with sensors in the asphalt, in all of the traffic lights and streetlights. We have sensors that tell you when it is raining, when it is too hot, sensors counting cars on the roads to help adjust traffic speed from real-world data almost in a live feed. Maybe even parking sensors in parking lots that can tell the operator exactly how many spaces are empty and how long people are staying (not just the total numbers but a visual indication of which parks in the complex are free at any given time.

Then we can start to consider the actual vehicles on the roads, especially as they start to become more autonomous and communicating with each other in a sort of ecosystem that will I believe in the end erase the need for traffic lights and traffic signs as the cars will just negotiate and work around each other to achieve the best results. They won't crash into each other just time the pass perfectly and continue on their own paths. They will be more efficient and will probably reduce the level of crashes (if not eradicate) on our roads as humans will not be in control anymore.

This might take some time before we let go of full control but it is certainly a future that is only a matter of time (personally I think It will ruin the pleasure and art of driving – I love cruising down a highway in an old school classic car – V8 purring and some great music – but we will all have to get with the times as this future is forced upon us).

Maybe we won't be on the roads and some of the flying autonomous vehicle start-ups will all get to the stage in which they will have an affordable and reliable option for everyone (to be honest I am a little excited about this one – I will be happy to sign up for one of these especially if it could make interstate travel faster and more readily available at minimum cost) I believe that this will reach the masses when one of these would cost the price of say a good quality SUV in today's terms (Happy to test one out if anyone of my readers is one of those start-ups mentioned above, pretty please J).

Okay so I admit, I got a little side-tracked but if all of a sudden there was thousands of these flying vehicles buzzing around our skies and possibly landing on the tops of our city buildings (instant image of "the Jetson's" popped into my head) they would likely all be communicating and transmitting a massive amount of data through this ecosystem, we call the Internet of Things.

What about our homes, we will have fridges, tv's, lights, door locks, maybe sensors in our grass or in the gardens all connected together telling us how our home is what we are doing and probably starting to predict what we will need and just buying it for us (I think AI purchasing will only be a matter of time if it is not already being used by some of those smart fridges you can see in homes these days).

Now as far as the IoT ecosystem goes, I think that the best way to think about what will be is that within a few years everything will have some sort of sensor with it, packaging at the shops will probably have sensors to help us know the most accurate "Best before Date", shoes, possibly even our clothes so we can monitor our temperature and other health factors constantly. IoT will quite literally be everywhere and I do not think we could stop it now even if we wanted too but there is a massive problem that many do not even consider when creating these devices/sensors. Can you guess what that is?

They are all going to be HACKED. (Probably not by me but they will be) This is not a prediction, it is already happening now. If you look at many IoT devices security is not even really a consideration, if we are lucky some may have a half-hearted attempt at adding some basic protections (which are probably little better than a device without it). This is not just about fridges or fish tanks anymore these devices will infiltrate our entire lives, monitor our health and could have a dramatic effect if taken over by a malicious actor.

They could lock down cities, cause complete gridlocks or just grind our lives to a halt. What about when malicious actors collect an army of IoT devices and use them for targeted attacks on our networks or critical infrastructure. Combine a group of a few billion IoT devices and it could be unstoppable. It's actually mind-boggling just thinking what could be capable if such a large-scale IoT zombie network could be built, could we really do much to withstand an attack of such magnitude.

We need to fix the current attitude that mass-produced sensors or other IoT devices do not need security. I know that many IoT sensors are low-cost devices and it will inhibit the use by adding security to them because of its increased cost to develop but it is important. Charge a bit more for these devices, the benefit of improved security will outweigh the possible $1-$2 increase in cost.

If we do not as a community come together and take control of the Wild West that we call IoT in today's environment, we will have no control. If that is the path, we are going to take, we might as well just pull out our firewalls and other security protections and just let the malicious actors in. if there are no secrets then we do not really need any protection for it (Please don't take that option).

There are already some organisations out there that are already working on this problem like IoTSec Australia but the cogs of change are very slow and we need to find a way to move this along at a speed that can cope with the avalanche of change that we are seeing with IoT.

So how do we protect these devices and make them more secure I don't have that answer but it is clear that if we don't find a way we will have a pretty dark future that is for certain, so let's pull together and make a plan to fix this nightmare so that IoT will be something that is beneficial for us all and not the reason for our downfall. Bring on some government regulations for IoT (I never thought that would be something I would encourage) and force a change that will be a benefit for us all.

Chapter 26 - Mentoring - A way to resolve the skills gap

Over the last few years, I have listened to many presentations, read chapters and white papers on the cybersecurity skills gap. Now I want to be completely honest here, I don't believe the problem is as big as it is being made out to be. Let's wait a few seconds for everyone to jump up and down and say I am crazy and don't know what I am talking about. I will just wait a little bit more for you to calm down and then I will explain why I think this is the case.

So, let's look at the issue we have, currently, there is say twice as many security jobs as there are professionals to fill them (Don't worry about the amount I am just hypothesizing here and guessed at the number). This situation is going to get worse as far as all the discussions around this go, so let's say we will only have 25% of the required professionals to fill the security positions in 5 years' time. Okay that does sound bad and we could just continue the doom and gloom angle but that's not what I want to talk about.

I want to look at this another way, everywhere I go I see amazing, smart individuals who want to get into cyber security or its related off-shoots, but they just keep getting knocked down at every approach. They have all the basic soft skills – inquisitive, keen, continuous learners who love to get in and figure out how things work but no one will even look at them because they don't have blah amount of experience or blah certifications. None of the Blah really matters for entry-level jobs but companies and hiring managers will not give them a chance.

Now, this isn't all about the companies or hiring managers either that is only a small piece of this problem. In my opinion, **WE** in security are a big part of this problem.

We have made these expectations of what skills or certifications someone needs and how many years of servitude are required before we are respected by our peers. I get that in senior positions it should be people who have the skills and qualifications to do those jobs but for entry-level positions which I am sure we have all been in some way or another this should not be a requirement. We need to drop the ego trips and look at what we really need in an entry-level position. Personality fits and the basic skills they will need for the job.

Security skills are learnt, we are not born with the knowledge and anyone can be taught. Yes, as in everything, some people are just naturals and will blow everyone else out of the water with their abilities and that's great. We should surround ourselves with people like them and learn everything we can. However, even naturals learn how to hack or code or whatever skill makes them so ore inspiring so why do we shut down anyone who wants to get into our industry, they won't all be superstars but that's okay too.

We need diversity in our industry if we are ever going to have a chance at winning this war we are fighting with the cybercriminals. Also, by diversity, I don't mean sex, race or any of that. None of that should even matter or be a consideration to us we should be looking more at the diversity in skills both in life and professional careers that could make our industry better at what we are trying to achieve. I think this type of diversity is what is needed.

So, if what I am saying is correct and we as the professionals in the industry are a big part of the problem shouldn't we do our part in fixing it? I think we should. Last week I was asked to join Cyber Century Mentoring as one of the Australian executive team. CCM was co-founded by Lana Tosic (New Zealand), and Amanda-Jane Turner (Australia), who both saw a need for quality mentorship to support those working or wishing to work in the many diverse roles that make up cybersecurity and cybercrime prevention. CCM is trans-Atlantic, and at this stage is concentrated in Australia and New Zealand.

The New Zealand Exec team is led by co-founder and National Director (New Zealand), Lana Tosic. Exec Team NZ TBC. The Australian Exec team consists of Amanda-Jane Turner as co-founder and National Director (Australia), Kristine Sihto as Senior Editor and Comms Manager and Craig Ford as Programme Developer and Outreach Manager (Australia) – Obviously, I accepted the invitation to join the team.

CCM is currently in what I would call the start-up stage of a volunteer initiative and is not yet registered as a not-for-profit in either New

Zealand or Australia. We are hoping to establish our presence and support of mentoring and we aim to be formally recognised as a not-for-profit association within two years. It is important to us that we start supporting mentoring as soon as we can instead of waiting for all the formal structures before we commence helping the community.

So why did I agree to join the volunteer initiative? It's simple really and to be honest, is probably the same reason Mandy and Lana started the initiative in the first place. I want to make a difference in our society and by mentoring or helping matchup great mentor/mentee I think we can help reduce the roadblock to entering this on occasion amazing industry as well as to help share our knowledge in a way that is both rewarding for ourselves and beneficial to the individuals wanting to get in.

I truly believe that by mentoring the up and coming talent from whichever industry or background they choose to come from we will all learn a lot from their varied experiences and help to educate a strong constant flow of new security professionals that will not only be able to hit the ground running but have the support they need to stay in the industry and succeed. It's a simple solution and I feel a great method to really make a difference to the apparent skills gap that is balancing on the cliff's edge. You never know we may even win the security war that is constantly waged with malicious actors/cybercriminals to protect our networks and critical infrastructure; if we can succeed in breathing new life to the industry.

Chapter 27 - MSSPs are just after their next sale

Before I start I want to be clear, I work for an MSP/MSSP in Brisbane, so before all of the rest of you, MSSP's and MSP's throw me under the proverbial bus read my chapter and understand what the point I am trying to get across here first. Then if you still want to do that then go for it.

I am a regular attendee to security events and meetups as I feel that I need to learn as much as I can from any source I can, it makes me a better person by always looking for the learning opportunity in every situation or interaction. Over the last few months, I have started to pick up something that troubles me somewhat though, a trend in response to the general question in professional socialising events or meetups.

Where do you work? Or What do you do for a living? Simple questions that we all get asked in many different situations but when I answer that I work for an MSSP (Managed Security Service Provider), I get a sort of negative reaction from some of the people I talk too. I thought this was a little strange, so I started to ask why this was so? (To be honest, I think I made a few people a little uncomfortable with that question, but I think it was necessary to get to the bottom of the cause). After some encouragement from me, I started to get a similar answer to my enquiry and it was basically this "MSSPs are just after their next sale, they don't really want to help. They just want to sell the next shiny thing in security and make their sales quota".

The first time I heard this response I was a little surprised, in the organisation I work for that is not what it is all about at all. We got into the security arena as we were concerned about our customer's businesses and the effect that a large-scale breach could cause. That's why I was hired in the first place to make things as secure as we could. So, as you can imagine I was a bit thrown back by that statement, but I then got the same response from a couple of other people I talked to about this.

This was what people thought about MSSPs and it made me a little angry if I am being honest. I have since done some more research and there is a trend where MSSPs keep just pushing solutions down the throats of their customers and potential customers to meet ridiculous targets set by their organisation's management. This is not what managed security services is supposed to be about.

I believe that managed security services are supposed to be about much more than sales, it is supposed to be about giving businesses who do not have a need or the resources to put on a full-time security team on, access to one and for that team to make that business more secure. We are supposed to give them access to skills and resources normally only available to large enterprises with large budgets and even sometimes be an extension to those security teams when the need arises.

This scenario is made possible as we spread the costs of training our teams in high demand areas that are hard to come by but make it a much more palatable cost than doing it individually in each organisation. It is more cost effective and it is easier to get access to different skill sets as needed. This model allows MSSP's to offer enterprise-grade services to small businesses not just the bigger end of town.

Now this idea that businesses need to just throw money at security and the problem will go away is ridiculous and this needs to stop **NOW**. Don't get me wrong sometimes you do need to throw money at a problem to get it under control but that should not be the standard response for every security issue. We as MSSP's need to stop and talk to the organisations and really listen to how they operate on a day to day basis. Consider how security is handled and firstly change behaviour so that a business can help protect themselves from mistakes that could be easily solved by just making them aware of the underlying risks that are accompanied by an action or process.

Teach the clients to be better and do better before throwing money at the problems, this on many occasions just Band-Aids the problem and in some cases doesn't actually do anything at all to help. Just to put everything out on the table so to speak, yes MSSPs are businesses and need to make money in order for us to still exist and provide the described services to our clients but do it in a professional way and don't gouge your clients. Charge reasonable prices for good quality services, it's pretty simple and something I feel we do at Davichi, but this isn't just about Davichi it is about the industry as a whole.

So, all you MSSPs out there reading this, pull your head in and do the right thing by your clients and the rest of us in the industry trying to do the right thing by our clients. We can all still make a good living out of providing these services, but it will hopefully stop me having to have these awkward conversations with random strangers.

Chapter 28 - Hacker Mules - Barely above the poverty line

During 2018 cybercrime generated around $1.5 Trillion Dollars (I just pictured a Dr Evil scene with his pinkie pose and laugh he had in the Austin powers movie from the late '90s in my head) for the underworld economy and possibly APT's. Over half of this was generated via dark web markets where contraband is sold or bartered ($860 billion). You could then break it down further with Trade secret & IP theft ($500 Billion), data trading ($160 Billion), Crime-ware/Cybercrime-As-A-Service ($1.6 billion), Ransomware ($1 Billion). No matter how you look at it, that is a huge financial burden on the global economy.

Now I have joked around in a few of my previous chapters that malicious actors are buying themselves nice new Lamborghinis (or any other ridiculously priced supercar you want to insert here – Admittedly I am only saying ridiculously priced because there is no way that I can afford one of these ever). They would buy them in several different colours, so they can choose which one to drive based on their mood on any given day. We would not want them to have to drive the same car for two days in a row, now would we?

Okay, I got a little side-tracked, but you get what I am trying to get at, is it a realistic picture of how most malicious actors would live? After some comments on my "Day in the life of a hacker" chapter, I have done some further research into that exact fact and honestly most are not living like this at all. The people who are doing the malicious work are what we would call mules in the industry. These are very low-level members of cybercrime gangs or Criminal syndicates in general, that are used by malicious actors to do all the work. This is a particularly common scenario in Russia and Ukraine, many of these are not even willing volunteers but are forced to do this work for what could be considered very poor work conditions and remuneration. Certainly, won't be getting a dental plan from this employer let alone a nice shiny new Lambo...

Many of them do it under personal threat/duress to them and their family's or they have no other source of income to support them and their families. So even though they are the ones doing the crime they are not the ones cashing in on the money train that is cybercrime, it is their bosses and bosses, bosses or even their bosses, bosses, bosses (that's a bit of a mouthful) but you get what I am trying to say. These mules are usually the ones who go to jail for the crimes, not the higher-ups, who actually get all of the money.

Is this the more realistic picture of how cybercrime is in the real world? To some extent, yes it probably is but there would still be some out there cashing in on these opportunities not linked to crime gangs or APT's that are our government-sponsored teams (in some countries they may actually be the same entities – now that is the scarier scenario and I can think of a few that could fit in that category). I think it is good to look at all scenarios, so we can get a better understanding of how this all works and truly try to consider what options we have to fight these groups.

It is a sad cycle really that is costing us all too much money but what can we do to solve this problem? It is a hard problem to resolve and what we are doing is not really working. We need to find a better approach, or we will never be able to win the cyber war (yes, it is a cyber-war) that is taking place all around us between the good guys and the bad guys (It's getting harder to tell who is who these days) but nonetheless that is the battle that is being waged.

Amanda-Jane Turner - the Brisbane AISA branch executive, made a very good point during a presentation at the BrisSEC19 conference that "Technology will not be the solution to cybercrime" and I think she is correct, we need to look at how we can approach this issue from a completely different angle, the solution may come from a totally different insight or industry and it may be so obvious that when we do have someone suggest it, we will all think "why didn't we think of that".

Someone reading this chapter may have the solution we are looking for; we just have to get that idea into the mainstream and action it. Let us stop the talk and make a plan to get this problem stamped out. If we do not find a solution to this growing issue, we may lose control altogether and the internet that is starting to look more like the Wild West may become so unsafe that it will lose the benefit that it was envisioned for.

If you have any ideas let's discuss it and consider its implications if it has potential, let's find a way to get it moving towards its end goal, if that fails, let's just try again and again until we gain back some sort of balance to this fight.

Chapter 29 - Collaboration -
Entering the lion's den

This is a strange topic as the idea of collaboration in the security industry is basically that direct competitors come together and openly share how they do things, how they secure their systems, what works and what doesn't. A couple of months ago now I was invited to join such a group and go along to one of their information sharing meetings.

If you would listen to all of the negative talks about this it would mean that I would be entering the lion's den so to speak, walking into the territory of my direct competition and to make it worse again I would not be just sitting down with that one competitor but in fact a room full of around 8-10 different competitors. In this situation, you could be forgiven for imagining a lynching party waiting to grab you when you entered the board room or maybe a wild west saloon in which all the cowboys in the bar with itchy trigger fingers, are standing ready to let guns a blazing at any sudden movement.

It is true that you could consider all these situations to have a similarly fine line of stability, obviously, a meeting full of competitors is not quite that dramatic but its success does still balance on a knife-edge. A wrong step could set the fragile state to collapse and no open honest sharing will take place. Andrew the guy that has organised the group and brought us all together has done so out true belief in collaboration and it is a noble cause that all of us attending the event believe in (at least enough to get us to the meeting anyway). So, the day came about, and I was heading to the location in Brisbane. I was a little early but took the elevator and made my way to the reception area I could see on my right.

No one was at the reception desk as it was after hours and most people had already left the office. I looked over to the reception seating area and I could see two other people sitting down casually chatting, so I walked over and asked if they were here for the same meeting. They both gave me a once over and one of them indicated that they were. I took a seat while we waited for the organiser to come out and take us through the lion's den to our allocated meeting room. To be honest the whole experience was thus far very easy going with no sign of awkwardness or discord from any of the attendees. A glimmer of hope that this could actually work.

The meeting started with some general discussion points being displayed on the projector from our host but after a beer or two, many of the attendees relaxed and information was flowing between all, with no hint of anyone holding back. Obviously, we weren't discussing any company secrets or anything that would be detrimental for direct competitors to share but what systems we have tried, what we are using, what is working well for us and what we think could be better.

We openly discussed a lot of information about how we do training for security to help generate better awareness, what sort of buy-in we are getting from senior management and staff. It was really good. The process was actually working as wanted, I am a huge advocate for collaboration and working together to solve our security problems (I am sure this works for other issues as well but my focus is just fixing security for now) but I have always received push back from peers when this was discussed or mentioned, so to see it being done successfully was a truly great feeling.

After a night of pizza, beer and some really great sharing of experiences, I think everyone in the room left with some insight they didn't enter with which was the whole idea. We all went our separate ways and we are due to have another gathering again soon and I would have to say that in its current format the group will continue to be a success.

The problem is, could this type of collaboration be expanded to be a collaboration with a whole industry or let's say Brisbane as a whole for the security industry? To be honest I think it will be a challenge but if this group has shown me one thing is that it is possible if we believe that it is. Let's drop our walls and try to find ways we can work together more and really collaborate.

If we can really do this, it will be a benefit for us all. We may in time be able to claw back control of this cyber war that is being waged. We have APT's, cyber-criminal gangs, financial or ego-driven malicious actors acting alone or in small teams. They are all collaborating better than we are, but we can't seem to just get over our ego's and come together for the overall benefit to society and in turn our customers. They are why we are still in business After all, so shouldn't we do everything we can to protect them?

So, let's put the differences aside and find a way to make this work for us all or the alternative is to let it all crash and burn as alone we can't win this fight. Think about it you know I am right.

Chapter 30 - Medical IoT - The future of assassination

I want to do something a little different with this chapter, I want to take you down the rabbit hole that is IoT and the dark side that is lurking around the corner that I know is going to bare its ugly head at some point in our future. I am going to create a bit of a scene for us to imagine that will show us a situation that is soon to become commonplace in our world.

Picture this, it is 2022. A businessman is working in the office late at night, he is reviewing a tender for a large government contract to build defence equipment of some kind. This contract will make his company and he is really putting in the effort for this tender as he knows how critical it is that he gets this right. As he starts to wrap up the tender documents he decides to have a drink of scotch as a job well done, he is happy with how things have come together and believes his company has a really great chance at winning.

He pours his glass and looks out his window over the cityscape in front of him. Suddenly he felt a sharp jab in his chest, he instantly grabs at his chest, he can't catch his breath. The look on his face is fear he knows this could be his end, whack, another quick jab in his chest, followed by another and then another. His heart stops, his glass falls from his hand spilling its contents and he falls to the ground **DEAD**.

In less than 30 seconds, everything changed. If the body is autopsied, it will show a heart attack with his pacemaker shocking him as designed, to help save his life, but it just wasn't enough to save him. Let's give him a name, let's call him Jim. This seems like a really sad situation in which the stress of the tender had become too much for Jim to handle and he succumbs to the stress with a heart attack, but this situation is much more sinister than it appears, it is the first assassination conducted using a medical IoT device.

Let's put a bit more background to this, Jim two years ago had a pacemaker installed due to irregular heart palpitations, the pacemaker was a newly developed internet connected device that would allow it to be connected to Jim's phone and send data back to his doctor about his condition. This could make it easier for him to predict issues, allowing for the treatment to be more accurate and improve Jim's chances of a long life with the condition. The device was also designed to receive updates via this connection to ensure that it had the latest programming and efficiencies.

The digital assassin used this connection to update the pacemaker with their own modified version of the software that had a hidden remote-control function. This function was used to zap Jim's heart with the full power of available electric shock to his heart multiple times until the heart was stopped. The device was then instructed to restore the previous version of the software, erasing all traces that it had been changed, although it will just look like a heart attack with no signs of foul play. The first IoT assassination has taken place with no one any the wiser.

Jim's competitor for government contract was very worried about the tender and new that Jim was the frontrunner to win, so several weeks earlier he reached out to a criminal organisation to try and hire a hitman to take out his competition. He wanted it to look like an accident or something that wasn't suspicious but if it could be done he knew that Jim's business could not take on the contract without him, leaving them to pick it up.

A murder has been carried out with a device that was designed to save our lives, imagine what else could be capable as these medical devices get smarter and more connected? They will have the ability to save or take our lives in an instant. It really is a scary thought, with the current lack of security in IoT devices this could be a terrifying scenario for us all moving forward.

I know I went a little down the Hollywood rabbit hole in laying out this scenario, but I wanted to generate a clear picture of how medical IoT devices could be used in the future and a scenario like this could possibly take place. There are already smart pacemakers that could be manipulated by a malicious actor. Security researchers at Black Hat conference – August 2018 - Billy Rios and Jonathan Butts demonstrated such an attack. They used the pace makers programmer which controls the electrical impulses that the pacemaker sends to regulate a patient's heartbeat. They indicated that 33,000 of these programmers are in use called a CareLink 2090.

This shows us that not only could this situation occur in the future that it may have already occurred without anyone being any wiser of its occurrence, a perfect murder if there is such a thing. One concern about this situation is Medtronic (the company who makes the products) who had been notified of the security issue by the researchers were more concerned about protecting their company brand then their patients who are at risk.

Pacemakers are not alone in this scenario though, what about insulin pumps which automatically dispense the dosages for patients. A quick and simple way to kill off a target without ever actually touching the victim and with possibly no way of ever knowing who conducted the attack.

We could also see situations in which victims are held hostage by a malicious actor with a demand for payment or the device will be used to kill them. I am sure most people will pay anything that was asked to ensure they are not killed by a cyber assassin, it's like the ransomware model on an all new level.

That is a thought that will keep us up at night, so let's get our crap together and find a way to ensure that these devices are not vulnerable to remote attacks (I know that is much easier said than done but we need to find a way). We need to ensure that all medical IoT devices are developed with the highest level of security in mind, it doesn't matter if that means they cost three times as much to manufacture/develop the cost of failure will be our lives if we get it wrong.

I seem to come up with all of the dark gloom and doom scenarios with these devices, assassination by a pacemaker, in a previous article I depicted a lovely scenario in which a smart car could be used to assassinate or kidnap one of its passengers. Death by insulin overdose and so many more scenarios. Sounds like we have a terrifying future to look forward to unless we can get the security right and to be completely honest about our history of success doesn't fill me with too much confidence but doesn't mean we shouldn't try.

Hopefully one of my readers has the solution that will change this fate from ever occurring in the future and will develop an IoT security standard that will keep us all safe for millennia to come.

Chapter 31 - Office 365 - Malicious actors using your account to scam your contacts

Office 365 is a name that is widely known and is the preferred email hosting platform for many organisations in Australia (unless you're a Google fan. It is like the age-old Ford and Holden rivalry – I am in the Ford camp here.). All rivalries aside though both platforms have their upsides and downsides it just comes down to which one works better for you and your business. I personally like the office 365 platform though as most SMB's get a really good quality system at a price that is very reasonable plus it just works with no real changes to how they used to use it with the old in-house servers they would have nearly all had before. It just makes sense to them which is great, but this isn't a sales pitch for office 365, no this is to tell you how it really is and set some things straight with no stupid jargon that is just used to confuse people.

So, let's put together a scenario here of an incident that I have seen on at least ten different occasions over the last six months. We get a call from an organisation who is hosted on office 365 with emails, software etc as is pretty normal. They have staff located across several locations or a mobile workforce and they all connect into the office 365 for emails and possible some sort of data sharing.

They have probably been on the platform for 12 months or more and it has been working well for them. Sounds like most normal organisations on 365 or google hosting, Right? Yeah, it does. Now we received a call because one of the staff has been getting strange bounce backs in emails for emails that they have not even sent in the first place. Alarm bells are starting to ring, this sounds like an email account compromise.

First things first reset the password, immediately. Just to be on the safe side. Doesn't matter if it turns out to be something else that is the cause, it is safer to just reset it and cut access to the account if it is, in fact, a breached account. Export all the logs for review at a later point and then check the rules in office 365 web portal for that user I bet you in most cases that you will find a rule redirecting emails with "invoice" or "payment" or "account" in the subject into deleted items, RSS feeds or a random folder created hidden down under folders you already have configured.

The malicious actor will be looking over all these emails, changing the details and then putting them back in your inbox as nothing had happened. They will change account details and invoice amounts just for starters. The malicious actors will then usually move onto sending sometimes crude and vulgar emails to all of your contacts or some sort of scam to get you to open an infected document or change account details for payments.

These contacts know who you are and have dealt with you before. They are not suspicious of your email when they come through so will most likely click on whatever is sent through. It just takes that one click and the malicious actor has another victim. These emails are less likely to be picked up by email protections as they are from legitimate users who have no record or history of email abuse leaving this scenario to continue to move from victim to victim.

I have honestly seen this time and time again over the last six months. The problem is certainly getting worse and some awareness of this technique needs to be generated so that we can slow down the flood of victims. One simple change that is really very easily configured can make such a large difference in keeping your office 365 account secure and in turn all of your unsuspecting victims. **TURN ON TWO-FACTOR AUTHENTICATION**. It is pretty simple and is turned on in the admin portal for companies 365 account. It is just an option that is turned on by a tick box, nothing scary at all. Once that is done each user just logs into their office 365 account and will be prompted to set up the two-factor authentication.

They will have some options with using an authenticator app or text message via mobile. I recommend using the authenticator app over the text option (some prefer the text option as it is simpler and doesn't require another app being installed on your phone) as this will remove the number porting method to bypass this two factor method (you will be surprised how easy it is to get a number ported these days – even though it isn't meant to be that easy). This method would basically take ownership of your number and send text verification to your number now in their control.

If the authenticator app is used the malicious actor needs to have your device and be able to unlock it before the authentication can occur. Yes, it is still possible to achieve this, but the risks are greatly reduced to the owner of the account in this scenario. The idea is to try and make it so hard to get access to your account that it just irritates the malicious actor enough to just say it isn't worth it and then move on to the next target. It truly is that simple.

So, do yourself a favour turn 2FA on in office 365, if you suspect something suspicious is happening with your account reset your password and either investigate it further or find someone that can help you find what is happening.

Simple directions but very effective.

Chapter 32 - Social media and its hidden threats

Social media, where do I start. Okay let's start by saying that social media has taken over our lives especially with the millennials and all of the following generations, we live and breathe it, we go out to dinner and need to take a picture of our lunch before we can even eat (even if it takes 10 minutes to get that exact shot I want and my food is now cold), any time we travel or just go to the shops we feel the need to check in and tell our 10 followers what we are doing almost to the extreme where we tell them the last time we went to the bathroom (seriously anyone I know does this, even if you are my parents and I will unfriend you – no exclusions).

In today's society people share way too many details about themselves and their loved ones. **STOP THIS NOW**. No one needs to know everything you buy, every time you do anything or are just leaving your house. I don't need to know your dog's name, your old school photos, children's names and birthdays of your whole family as you post about every single one of them. I get some of you like to share your experiences and that is fine, even I use social media, but why I use it is probably a completely different reason that you do. I will give you a bit of insight into why I use social media and you may reconsider the way you currently do.

I am a HACKER (also known as a penetration tester or security engineer but the name doesn't matter), part of my job is to look for vulnerabilities in client systems and exploit them. Find vulnerabilities and weaknesses that will provide me with a way in, these could be systems, processes or even humans. If you share too much on social media, you are waving a red flag at a bull known as hackers. Yes, some of them are like me and are trying to better protect our clients or business but there are probably more malicious actors (bad hacker types, cybercriminals and so on) then there is of us good guys/girls.

What you all need to understand is that these profiles are all public, anyone from anywhere around the world can look over everything you have ever posted, and you will never even know they did it. I know many of you are probably thinking so, what does it matter if I have nothing to hide. If I wasn't willing to share the information, it wouldn't have been posted on my social media. The part that you need to understand about this is it isn't about any one particular post, it is about all your information that you are sharing.

When I want to gather information on a target company I look for all of the staff on LinkedIn and Facebook to learn all I can about them. To put together information that will help me guess your account passwords because you have probably already given it to me in pieces on your different pages. Let's see what you might have used, pet's names, favourite movie or foods, children's names and birthdays, partners DOB or anniversary. The list keeps going and going. All of this can be easily collated into a file that can be used to attack your accounts and let me straight into the company systems. That's not hacking you say, but it is I am just letting you save me lots of time and headaches breaking in the hard way.

It doesn't end there though either, all of this information will be perfect for me to conduct a social engineering attack against your organisation. I know a lot of information that could help me appear that I know you well, how is your son's basketball going? Did John pass his exams? I could have a great phone conversation with one of your team to make it look like I am who I am pretending to be.

How about we talk about your latest holiday to QLD? I could go on for hours on how the information that we all share on social media is used and abused by malicious actors (and us good guys on occasion) to succeed to gain access to your accounts. In my case I would just go after your company account as that is my goal, but malicious actors will go after your bank accounts (that would be a great surprise when you try to pay for the petrol you just fill your car with when all your funds have been transferred out to some nice offshore bank account). They will take control of your emails, social media and any other accounts they can get access too then use those to scam your friends, so they can then clean out their accounts as well.

Social engineering can be done without this information but don't make it easy for someone by giving them a treasure trove of ammunition against you without even doing anything to get it except look you up online. Sounds like an easy payday to me. This isn't the only threat that you face by oversharing, let us look at it from some other angles.

What about if you wanted to stalk someone, I could see everything you are doing know where you normally go, who your friends are even where you live (people should never share their full address on social media – if you have, take it off please). Maybe a burglar who has been staking out your home to clean out, gets your name from mail in your mailbox then looks up online to see your normal activities and you have posted that you are away for three weeks in Bali or New Zealand. You may as well have just left your doors unlocked as you won't have anything left when you arrive home from your most amazing trip ever that you couldn't stop posting about.

So I hope you all understand what it is I am trying to get at here, STOP OVERSHARING on social media, it will be the best thing you ever do, both to help reduce your risks of being a cybercrime victim and to be honest many of your followers/friend will probably thank you as well as it can be irritating when you have one of those friends who share more than any of us really want to know.

Take my advice or don't but at least know your risks than when your accounts are all breached you have no one else to blame but you're serious oversharing.

Chapter 33 - True diversity in the cyber industry

Diversity in the IT and security industry has been a topic that has been around for quite a long time and it is generally just focused on trying to get equality for women in the field. This is great, I think some of these initiatives have brought in some great talent to the industry with some very talented. Women that may never have chosen to join the security industry if it hadn't been for these programs, but I think this focus is to singular and I want to explain why.

So, before I do please don't jump to conclusions before reading this chapter all the way through, I want women and men to have the best opportunities possible and I am not a chauvinist or anything along those lines (well I personally don't think I am), I have a beautiful little girl and want to ensure that she is treated as an equal and is given every opportunity to achieve her dreams. So now let me explain what I am trying to get across about true diversity in cybersecurity.

True diversity is simple as far as I am concerned, it should not matter if you are man or women, gay or straight, Caucasian or Asian, have come from a background in accounting or IT. None of this matters at all. Not one bit and I want to explain this a bit further.

Let's look at the problems we have in security, we are faced with adversaries that are from all different backgrounds, experiences and completely different ways of thinking. So how can we truly defend ourselves if in this industry we do not strive to reach a similar level of diversity in our teams and the industry as a whole? It's simple, if we can't do this we won't have a chance of succeeding at all.

Another thing that troubles me is that although I believe that some of these initiatives that support women are great, some are not beneficial to any of us in the industry and are just used as a platform to trash the male gender. Does that really help any of us when men are not the issue as a gender, most men that I discuss this topic with are completely for mutual gender diversity it is only a small few that are just pigs to put it bluntly.

An entire gender should not be inflicted and demonised because of just a few, if the tables had turned I don't feel that women would be as obliging to this treatment that we as men need to just stand silent against, due to fear of the backlash and abuse that can be inflicted by any man or woman that speaks out against this as a movement.

Don't get me wrong I am not saying that there shouldn't be groups for women in security, in fact I encourage them, I would love to help as many women get into the industry as I can but what I want to make clear is, this can be done without trying to knock down the opposite sex. Let's forget the sexes all together and try to help the industry be truly diverse.

I want us instead to try and bring in new people to our industry that has a very different set of core skills then we would normally look for yes that would be both men and women but I am talking more about someone who has a psychology background, wouldn't that be a useful skill in trying to figure out how an attacker would think or maybe to be part of the red team trying to test security assets as they would be great for social engineering attacks don't you think?

What about a mechanical engineer, they may understand systems better than any IT person will ever be able to grasp. Then we could look at just country of origin or religious backgrounds, these differences will allow them to consider problems in a different light and that difference in how they had been brought up might just be the factor that allows them to determine a course of action that had not previously been even considered. That diversity in my opinion is the key to solving so many of our problems in security, so instead of being narrowly focused let's mix it up a little and encourage true diversity in this amazing industry we are lucky enough to work in.

Chapter 34 - Cyber Bullies and how to stamp them out

Cyberbullying is a big problem in our society with around one in five Australian teens being a victim. Wow, that is huge, we need to find a solution to this issue. Statistics from the Children's eSafety Commission show 19% of young people between the ages of 14 and 17 admit to being harassed or bullied online. Some of these teens never recover taking their own lives, what a devastating loss of young life that is unnecessary.

So, what is cyberbullying exactly? The kid's helpline website lists it as "Cyberbullying is using technology to bully or hurt someone else" they break it down even further with the following:

Cyberbullying can include:

- Sending/sharing nasty, hurtful or abusive messages or emails
- Humiliating others by posting/sharing embarrassing videos or images
- Spreading rumours or lies online
- Setting up fake online profiles
- Excluding others online
- Repeated harassment and threatening messages (cyberstalking)

These types of cyberbullying are sadly part of normal life for many young people all over the world and it is a distressing problem. Let's really look at the issue, back when I was at school (I graduated high school in 1999 – I know that makes me old now to all of these 14-17-year-old's) cyberbullying was not even a thing, we barely had the internet and computers in our lives let alone all of this social media that we can't live our lives without. I don't even believe we had a computer in our home, I didn't even buy my own until 2000-2001. Things have changed a little since then that's for sure.

If we had a fight with someone at school, it was usually a physical one and would normally last for a day or two and it would be over. If you were being bullied than at least when you left school for the day it was over until you arrived back at school again and you could avoid the bullies and in turn avoid being a victim of their tirades. The bullying was (even if it was still horrible) was isolated to school and could be kept away from your normal out of school lives.

However, in today's society that is not an option. Teenagers are always connected, always interacting via social media and if they are victims of bullying there is no escape, the perpetrators of this harassment can continue to bombard their victims on any platform and shame them publicly without any real recourse for their actions.

Sometimes parents don't even know what is happening to their children and that isn't entirely their fault. It can be hard to keep track of what our children do online, let alone if they are being bullied. Victims will in many cases hide this from their loved ones due to embarrassment possibly or just don't want to admit that they are not liked at school, it can be a real challenge for them.

What can be done about this issue though? There are some good resources on the kid's helpline website that can help parents deal with the issue including reporting the crime to the police but there needs to be a better way to deal with this. Why hasn't someone created an app that will help control our children's internet and social media access to help reduce or even prevent cyberbullying?

I think it should be possible to do this, an app that can be installed on mobile devices and computers by a parent. The app could then have control of the social media apps on the devices including web browsers. This control could allow the user to block out any content that is not suitable to be viewed by the user of the device but not just limited to this. It should then be able to block and educate users on bad behaviour when they are typing an instant message to someone and when they use language that is offensive or hurtful, the application could be designed to block the user from posting it or sending the messages.

It could stop snapchat from sending graphical or inappropriate materials (not sure how that would work but I am sure it's possible). Block Instagram content or anything else that could be deemed inappropriate. It could educate the user on their behaviour telling them that what they were about to do was inappropriate, why they should reconsider this type of behaviour and notify their parents of the incident so that parents of bullies know what their children are doing.

Don't we all think that whether our children are the bullies or the victims we would want to know about it? This type of behaviour can have severe consequences to developing, fragile young minds and often results in tragic loss of life. I know I would want to know if my child was responsible for bullying others so that I can ensure they understand the full scale and consequences of their actions. I would possibly search out families who have lost someone from bullying and ask them to help my child know what the true costs are. Yes, that may be at the extreme end of response but if it could save lives would we not want to educate our children no matter the side of this horrible behaviour they are on?

Look I get it would be tough to find out that our children are the bad ones in this tale but if they were and we could stop the behaviour before it gets out of hand, it's a much better result for everyone involved don't you think?

Someone must be able to create this app that could make this dream solution a true reality? Colin Anson of Pixevity has come up with a solution to give parents control of images taken or collected via schools and allows them to have a choice of whether or not they can be accessed by other parents and the school. Parents have the full control no exceptions. This is possible with a facial recognition capability in the platform that recognises your children and allows parents to decide on their privacy levels. This is a great looking platform that is started to be used in Australian schools and from what Colin indicated at BrisSEC19 they are also being utilised by overseas schools now as well. If Colin can find a solution to his problem with giving parents control of all images of their children, there needs to be a way that we can create a platform to curb bullying behaviour.

Now I am going to put the challenge out to all of my readers and hopefully anyone in their circles and their circles circles that one of you can make this a reality. I know it can be done and if anyone who thinks they can do it, reach out to me and tell me your idea of how to make it work. Together we can stop unnecessary loss of life, let's make a solution and get it on all our children's devices. I know some of you will say this is an invasion of their privacy, but shouldn't parents be able to control their children's activities to ensure they behave in a manner that both society and they deem suitable.

We can ensure that only parents have access to this control via some form of multifactor authentication to prevent the children/user getting around the control and any malicious actors getting control of the platforms for some nefarious reasons (well at least do our best to try and stop it anyway).

So, reach out to your contacts and ask them to reach out to theirs until we find someone who has a way to make this happen (and tell me about it so I can help in any way can) it will be a much better world because of it.

Chapter 35 - Is your mobile workforce secure?

In today's business environment it is almost guaranteed that your business has some type of mobile workforce. It may just be a couple of sales people that access your CRM when on the road or do you have remote workers who don't always come into the office to do their general duties? Maybe, you just have some staff that like to remote into your network after hours to catch up on the day's workload on occasion. If so I am sure you have a strong mobile workers policy implemented with the ability to lock down any of these devices remotely as needed? You don't, oh that could be costly...

I know you are sitting there going what do you mean that is going to be costly? These are remote workers, these machines are not connected to our corporate network, how can they be a security threat to our business? Well, I am glad you asked (okay maybe you didn't but go with me on this), let us look at the issue more closely and really understand the risks involved with remote and mobile workers.

Let's look at a sales rep – we can call her Jane. Jane has a work laptop that she takes with her in her daily travels visiting customer sites to check that orders and stock are meeting expectations and sometimes placing further orders for her clients while onsite. The laptop is a windows 7 machine about 5-6 years old now with an expired antivirus subscription on it (this is a trial licence that was on the laptop when purchased). The machine has automatic updates turned off and has never been looked over by the corporate IT team since it has been used by Jane. Jane uses this laptop to connect into the remote desktop servers hosted on the corporate network to access the CRM for the customer details or order information as required.

Jane also uses the laptop for company and private emails as well as personal internet browsing. Jane's children also use the laptop on occasion to do school work and assignments as well as play online games. Seems like a normal kind of usage scenario for most people's laptops, right? So, what is wrong with that you say? Let's look at the scenario a bit more granular with security in mind as we do.

Let's start with the laptop itself, it is 5-6 years old with Windows 7 as its operating system. Windows 7 is soon to be dropping support and will no longer receive any security or critical updates. The laptop has windows updates turned off and has never been updated. Please tell me you see an issue here? Over 5 years of not doing updates is a massive risk, in that time there could have been hundreds or even thousands of vulnerabilities found with patches released by Microsoft to resolve them.

What about the expired antivirus? If it was just a trial that had come with the laptop at its purchase it would have been for 30-90 days of coverage in most instances. This means the laptop has been unprotected for at least five years. This would mean that the laptop could be infected by any sort of malicious applications or viruses that could be collecting customer data or key logs or more. They could have used this machine to gain access to the corporate systems as it is used to remotely access it, is it not? You think you would know if you have viruses or other infection it is possible that you have no idea that your systems are infected especially if the malicious actor keeps activities to a very low level to avoid detection.

The fact that the laptop is also used by the other members of Jane's family for games or internet activity it could be an easy jump to think that someone could have downloaded a malicious application. It could have even occurred from some sort of phishing email as both personal and work emails are accessed on the laptop. This cold again allow the malicious actor to gain control of the corporate email account that Jane has, a stepping stone to gaining access to the corporate systems or just using your company's reputation to get unsuspecting contacts of Janes (All your hard-won customers) and spread their access even further or infect them all with a ransomware bug. I am sure all the contacts will love that gift from Jane and your company (yes if the malicious email comes from one of your sales reps they will blame your company for it – Not just Jane the sales rep).

So, do you still think that this machine is no threat to your company? I'm glad you are starting to come around to my way of thinking. So, what can you do about it? Lots of things but how about we start with the basics and then you can improve from there.

Firstly, you need to ensure that your company has some type of remote management capability to manage the laptop, including patch management so that you can ensure that the systems are as up to date as possible.

An antivirus solution should be provided on all systems that are used to access company systems, yes, I know this is an added cost but the small cost to add a few more licences will greatly improve the security of the machine, so definitely worth the small pain the cost will inflict compared to the possible cost to reputation/brand damage that a breach would cause.

Company emails should be restricted to the remote server login only and not on the local machine, multi factor authentication should be configured for emails and remote server logins as well as VPN connection. The systems should be isolated to only allow access from VPN connected machines or internal to the corporate site.

Beyond this, your company should develop a thorough teleworker policy that clearly outlines company expectations when it comes to its staff accessing the corporate systems remotely. This will help to ensure that all staff follow best practices and keep your network as safe as possible form threats.

Of course, there is more that can be done to protect these systems from threats but it is important to first embed the basics before trying to build advanced protections, many organisations make the mistake of trying to just jump to the more advanced options as they find the basics to be boring but please take my advice and do the boring stuff first you will be thankful for it later as in many cases it will be that foundation work that will, in the end, protect your networks from a major breach.

This chapter is a good demonstration of how simple things that we think couldn't really be a big threat to us, can once investigated turn out to be a fair high-risk threat that could be the cause of a major breach or incident, but the solution was simple changes that had just been ignored.

So, if you just learn this one thing, ensure that you put your head down and get your basics in order.

Chapter 36 - The blame game in security needs to stop

Security is a hard gig, honestly, it really is. We need to know about all the threats that haven't even come to be yet, protect the organisation with all the latest blinky light solutions as well as know the ins and outs of every application and its implications to your company. After that, you need to be available 24/7 for any incident that may suddenly just pop up no matter what plans or personal life you think you deserve. What about the endless hours of dredging through logs and alerts just to find that they are all false positives, but what happens when after two weeks of 14-hour days and you miss one instance that was a legitimate threat?

The organisation that you have almost killed yourself defending throws you under a proverbial bus just to save face. Does that sound fair to you? Hell, no it doesn't but it is happening all over the world right now. When breaches occur which honestly is every day now, these security professionals who are overworked and sometimes balancing on the edge of both a physical and mental break down put everything in their own lives to the wayside and do everything they can to protect their organisations from all forms of attack. They may even be successful at stopping the threat but that doesn't matter.

If it comes to the crunch and the organisation needs to blame someone for the issues the finger will point straight at the CISO and his security team. "It is their fault if they had done their jobs properly then this wouldn't have happened" but that is bull S#!t. It honestly is but this is truly a reality in the security world, it needs to stop.

Let's actually try to look at the pressure some of these teams are under and look at the growing adversaries we are all fighting against. This is a war we are losing and shooting all our soldiers for trying to fight the good fight is the stupidest thing I have ever heard of in my life. Yes, I get it sometimes it really is the fault of the CISO and maybe some of the security engineers but how about we change the default finger-pointing attitude that is so rampant in our industry and in the media.

Let's acknowledge that many times we are just outmatched and outgunned in this fight and all we can do to survive is put the fires out as best we can. On those occasions when it is their fault and it was negligence, okay I understand, throw them under that proverbial bus that just keeps rolling by, give them fines and whatever punishment is reasonable for the malpractice they committed.

However, don't destroy the lives of hard-working security professionals doing everything they can to save you, if they lost at no fault of their own then it's not really their vault. How about we help them out and if it turns out it was because they are missing a skill that may have helped them match the malicious actor's skills, why don't we train them instead of firing them for doing their best.

If it is their fault and they have just made an honest mistake, then that needs to be a lesson learned not to pack your things and leave. Honestly, anyone would think with the amount of security staff churning with all this blame floating around that we didn't have a skills shortage...

Oh, hang on wait, apparently, we do have a major skills shortage, that in the coming years is only going to blow out to massive proportions, but you want all of the potential new recruits to come into an industry that does not support their own, we just throw them aside and move on when it is convenient for us. This just seems like we have monkeys behind the wheel and no one is actually using their brains to really consider the consequences of this problem.

So how about we do something a little different, why don't we instead make a decision that as a company we are in this fight together. Support our staff and take an active part in this together. Don't get frustrated when the security team needs to take systems down for a 30 minute window in the middle of the night to patch systems, listen to them when they say that the risk of whatever thing you want to do is too high and at least consider that they are the ones working all night to save your systems instead of being at home with their families.

Be patient and ask them if you can help in any way, just offering could be enough to make their day just that little bit better. But most of all, if the house comes burning down around you don't just throw them under the bus to save face, stand tall with them and help put out the fires. This will be a better result for all if we acknowledge that the fires are spreading, and we need to all stand together or we have no chance of surviving.

Breaches are not an if anymore, it is a when (if it hasn't already happened and we just don't know about it yet) so let's stop all this stupid behaviour and do the right thing by our teams.

If we can actually do this, we might even look good to the potential new entrants in the industry and they will join us in the fight to eradicate cybercrime (or at least get some sort of control back). This will mean we will have happier security folk with better training because instead of throwing them into the garbage we will help make them better.

Security is a taught skill, we aren't born with it. So, don't have a tantrum when someone doesn't know everything we are all just human after all. Okay, I think that is enough of a rant, for now, just be better.

As usual, if you don't agree to tell me, I want to have an educated discussion about this, so speak up and tell me what you think. We don't need to agree, we just have to be open to other opinions, that is in my belief of how we will find the solution to all our problems.

Chapter 37 - Be good corporate citizens - The Insider Threat

In many of our organisations, the technology arena is almost like the Wild West with the so-called shadow IT and BYOD removing control of systems from the people who have been given the task to keep it safe. It really is getting harder and harder to keep any sort of control to this mayhem, but we need to find a way to wrangle it in or we will never take back the control we need to defend ourselves from threats that our businesses face on a day-to-day basis.

Let's start with shadow IT, it is sort of a buzz word that has propped up over the last ten years or so in which you have staff bringing in wireless access points or storage devices that have not been provided by the IT department and gone through the process of hardening to ensure they are safe to be used on the corporate network.

This was a huge problem for a long time until organisations started to adopt wireless networks as part of their own internal network structures as staff would want to have wireless access on their iPads or smartphones and their requests to have wireless added would be denied due to the belief that wireless networks were insecure (which they generally are pretty insecure – even the new encryption standard WPA3 is already broken and it's not even really adopted yet). Don't get me started on personal cloud storage in business, you see this everywhere and there is no real control of data flow in or out of organisations, that is all of your intellectual property being transferred out and you probably don't even know about it. I am serious, go check your logs it is almost guaranteed that at least one staff member is accessing cloud storage for one reason or another (yes it doesn't mean that it is malicious but you don't have any control of these accounts – that's a bad thing).

I know that it can be a pain to go through all the right channels to get something setup especially if you want wireless or cloud storage but we all need to just suck it up and deal with the bureaucracy that is IT in especially mid to large enterprises. Ensure that security is top of mind in all things, these configurations need to be done correctly to ensure they meet minimum security standards.

What about the BYOD nightmare? This is an interesting one and is kind of a run on from the shadow IT, I feel that IT departments just caved in and gave up on trying to stop staff bringing in their own devices. Users want all of the latest gadgets and they are always faster to adopt these technological changes, but it has the same issues as the shadow IT except now they tell us that they are connecting the devices to the networks (I guess that is a little improvement) and using work emails on them. We need to ensure that we get the right permissions from users of these devices to allow organisation provided protections to be added to their devices and for remote wipe capabilities to be enabled to ensure that if a device is lost then we can just encrypt all of the data or brick the device so that nothing can be retrieved from it.

That allows some kind of control of these systems but if we can do this at minimum, we can start to get the control back and keep our systems safe. BYOD is a challenge that is for sure and it is hard to force users to allow such controls or security measures on devices that the company doesn't own but this can be done by making IT policy that If a user would like to access company resources than they need to agree to these controls before access is granted. This can be done as if multifactor authentication is on and you have device lockdown you can ensure that only authorised devices can access desired systems.

Insider threat goes deeper than shadow IT and BYOD though, with many risks that are not always malicious but can still put your organisation at high risk. Staff can fall victim to social engineering attacks by malicious actors, they can accidentally share sensitive data on the public domain or bring down entire systems whether accidentally or maliciously, but it can indeed happen and has in the past. We need to first educate staff on best practices to help prevent the unintentional instances as well as ensure that they feel comfortable to come to the security or IT team if something does happen so that the damage can be minimised.

This openness will be much better for the organisation moving forward than if they don't feel comfortable in approaching the team to get help fixing the issue as they will just try to cover it up with fear of severe retribution, accidents happen, and mistakes are made we need to just come together and try to make the best of a bad situation.

What about deliberate theft of data or systems sabotage of a hostile employee has someone been fired and is looking to bring forth revenge. Systems need to be in place before you are in a situation like this to ensure that you cannot just manage but detect an incident if it is occurring before it gets out of hand. You need to have a SIEM that enables you to track and detect unusual activity for users and send an alert to your security team so that it can be investigated. If it is a real instance of malicious activity action can be taken quickly and quietly to avoid further damage.

How can you react if you don't have vision into what activity is occurring on your systems, you can't, it's that simple. A siem alone won't resolve this problem either, your organisation needs to have processes in place to manage this threat before an employee is terminated. A high-level request or ticket needs to be generated and actioned at the correct time to close out any access before the damage can be done or data can be stolen. Just a few simple steps to ensure the protection of your organisation's data, be prepared you will be thankful for it later.

Chapter 38 - Fighting cybercrime is everyone's business

Cybercrime is a problem that costs the Australian economy up to 1 billion a year in direct costs alone according to ACIC (Australian Criminal Intelligence Commission), this is a ridiculous financial burden on many hard working companies who just want to do what they do best. Make or sell a product or service in which they are operating to perform. The sad truth is a high percentage of these organisations are not prepared for cybercrime as a threat, it is just not even on their radar of problems that need some of their limited time resources.

Especially if we look at small businesses with say up to 20 staff, these people do not have time or the financial backing to prepare for cyber threats or at least that is the opinion of many of these organisations. I am regularly told in my day job that security is not even something that the business considers. That is a little scary if you ask me but what they are saying is true they do not give it a thought and even when it is discussed with them many get that glazed over look in their eyes. I get it, security is my thing and I love talking about it and that is really not the case for many others (I get that same glazed over look when someone talks about sports – I am not really a sports guy) so I do understand that to some this is just white noise.

This is a little off topic but it's even scarier when these same organisations do not even have an antivirus solution on any of their systems and the few that do over half of them are expired. Now, antivirus protection is the bare minimum protection that you could get but if you don't even have that you are basically screwed no question. You are probably already infected with some sort of bug and are probably sharing all your data with a malicious actor. You could actually be in big trouble with this type of scenario as in Australia you need to notify all customers in an instance of a data breach and if you have cyber insurance you might as well burn the cover document as they won't be paying you anything with the severe neglect to your security responsibilities (and probably lied on the information form when getting it as it asks what protections you have in place).

Okay back to cybercrime now after my slight topic run away. If you ask businesses about cybercrime you will be told repeatedly by the leaders of those organisations that "That's an IT or Security problem" and they are partly correct it is a problem that these areas should be involved in but Cybercrime is everyone's business. What do I mean it is everyone's business, simple.

Cybercrime is a big problem and we all know that with the 1 billion direct costs stated above but to fight this we need more than a technology solution or just a security solution. We need to reach deep into all areas of our businesses and every citizen or visitor to our country. Heck, let us say everyone in the world needs to get involved (except the cyber criminals of course, as they are the ones costing us all this money and well they wouldn't want to help curb this back it might mean they can't do that overseas holiday this year).

If we can bring everyone into this fight and look at all solutions that could help improve our resilience to methods that cybercriminals use against us all we could half this figure or almost completely eradicate it (It's unlikely we could eradicate it completely but it doesn't hurt to try). There are so many things we could do if we have everyone on board, we can change businesses processes to double check and triple check financial change requests or payments, we can change the policies of how staff do certain processes or interact over the internet, require direct contact for customer details changes.

What about training our staff or doing a compulsory internets smarts program in our schools to help educate our teenagers, we could then move on to the elderly I am sure we can find ways to help educate and protect these individuals from threats. I will not accept that the problem is too big for us to be able to find an adequate solution to improve these results for everyone. How about a life be in it type government campaign to really drive home the risks for everyday Australians and once that message has been driven so deep into our subconscious we can then move beyond this with more targeted businesses campaigns that could really help turn the tide so to speak of cybercrime and its mounting pressure it is putting on the world's economy.

I know I am just throwing ideas around here but we need to do this and we need to debate what will be the best path we can take to achieve what we are aiming for. We can leave it there though we need to grab hold of that decision and find a way to get it happening to ensure that we do not just talk about it but knuckle down and get it done. That is the only way we will make a difference, try something if that doesn't work adjust and try something else until we can find a solution that can work. It may be five different plans that all come together that gets the result but if we do nothing I can guarantee you we will never succeed in finding that solution we need.

Chapter 39 - The attitude adjustment needed in security

Since 2013 I have been ingraining myself more and more into the security industry due to my desire to really dig deep and learn everything I could to make me a beneficial member of this industry. I didn't want to just learn the basics and continue a more IT orientated path in the field I really wanted to live it. I wanted to know everything that I could, I wanted to be able to truly help solve the pending avalanche of attacks on the systems of our world, to be part of the solution and not just stand by why it happens. A noble pursuit I thought at the time and I feel I have started to become a good security professional but during this endeavour, I found a dark tone to the industry that I want to pull out into the open and air out the dirty laundry so to speak.

During the last 6 years or so years it has become obvious that our industry has what I have heard called the "Rock star" syndrome. Basically, it is in most occasions really talented security folk (sometimes they don't even have any talent but still have the attitude) who act in my opinion like spoiled rock stars. They won't get out of bed for less than a million dollars (not the real figure but you get what I am saying), they won't start until 10 am, they will only work in a room that has unlimited coffee or energy drinks or sugary/salty snacks. They only want to work four hours a day and obviously, they won't do mundane tasks that they were not specifically hired for as that is beneath them.

Okay, so most security people are not like that but yes there are some that really are and honestly, these people are usually not worth the money they demand (so do yourself a favour and don't hire one of these or fire them if you already have one). Many of the ingrained security professionals that have been in the industry for ten or more years seem to have a pure arrogance to them some more than others and even if it is not quite to the extreme of the rock star syndrome, this level of arrogance, in my opinion, is bad for the industry.

Let me explain this a little more so you can understand why this is so bad. Let's look at my early experiences with many professionals in the industry back in 2013 when I was just starting to make a move towards a cyber career. I was keen (that is clearly an understatement), so I was reaching out to respected professionals and asking them for advice on what I should look at first, what training should I do. I personally feel I am a resilient guy and can handle rejection well (I like to think I have a thick skin) but almost every single person I reached out to in those early days flatly refused to provide any direction or advice.

I even had some of them tell me that if I had to ask for advice or where to start, I wasn't worthy of their time. They basically were closed of community that would not help anyone looking to gain entry to their ranks. Look, I get it, these people had been in the industry for years and they were the gods in their circles and I was a lowly old IT guy who wanted to learn how to do what they magically do. I had no reason that could make them believe that their time was well spent on pointing me in the right direction.

If I am being honest these responses are part of the reason I worked so hard to prove them all wrong and that works for some people but let's fast forward 6 years and every day someone else in our industry is complaining that there are thousands of unfilled positions in cybersecurity and that there is a huge shortfall of qualified security professional entering the field. Apparently, this skills shortage is getting bigger and bigger, but no one has a solution to fix the problem.

I have previously floated the idea of mentoring as a possible solution to the issues and I honestly think that mentoring if done right could really help make a difference in bringing in some great talent from all different types of backgrounds through, This could help to bring in different ways of thinking that could also help us as an industry solve this long-running security issues but this won't be possible unless the attitude in so many of our industry changes.

I know the industry is made up of some really skilled people who have spent lifetimes developing their skills but if all you do is gloat how great you are and then the next second complain that you can't fill the unicorn position that you have advertised because you firstly aren't willing to train anyone to do the job and the 10 year's experience that you are asking for with a beginner/starter position.

You act all high and mighty but your attitude towards someone who is just trying to find their way is pathetic and the fact that many in our industry just can't even see the fact that they are doing this to themselves is astounding, many are completely oblivious and stuck in their own warped sense of reality.

We need to reflect on our own industry and change the poisonous culture that is just driving potential members away. Let's embrace the fact that just because we already know a lot about security and are members of this great industry it doesn't make us rock stars, we weren't born with these Rockstar skills so let's not push people aside, let's take a few minutes of our time to consider their perspective and if we can give them one piece of advice that could help them or maybe refer them to someone who can that alone will generate a change for the positive.

Let's fix this arrogance problem that we have made and then focus on how we could be just better human beings. I think we will all be much better for it.

Chapter 40 - Never use free Wi-Fi

This is a conversation I find myself having almost every day and I believe that the message isn't really sinking in on the dangers of using public or free Wi-Fi. Using these free internet services opens your machine and your organisation to malware, credential theft, data/IP theft and so much more. I don't think the benefit of the free internet is worth all the risk it brings with it and to ensure that you all understand why I am going to describe a scenario in which I could use this situation to my benefit, I may even do two different scenarios just to drive the message home just a little more so we can get you all using a safer connection when working outside of your office.

Let's look at scenario one for the moment. You are a business person and you are flying from Brisbane to Melbourne, you have done this trip hundreds of times now as you are a regular traveller to all of the states and you are already through security and are now waiting on your flight to start boarding. It has been delayed for almost an hour which is going to mean that you will be cutting it fine for the first meeting you have in Melbourne. You intended to work on your presentation for that meeting in your company's office in Melbourne when you arrived but now you won't have time to do this and it is extremely important that you get this pitch right as this could be a really great customer if you can bring them onboard.

You decide that you are going to get your laptop out and do some work on your presentation while you wait for your plane. You get it out and look on your laptop for your presentation, but you have it stored on the company's cloud storage and save storage you don't sync files to your laptop. Your only option is to connect to the internet and download the file to your laptop, that way you can easily work on it on the plane as well. You click on the wifi button on your laptop to turn it on and you get some available options come up, one of them is "Brisbane Airport Free Wi-Fi" so you click on the available network and select connect (That was mistake one).

Now you open up a web browser and attempt to connect to the web portal for your company cloud storage, enter your corporate login details and download your files (Mistake number two). You then decide that since you have the file that you will just check if you had been paid your expenses claims for last week and logged into your bank account. You have a look through the details and then log out of your bank account (that was mistake number three). You then hear over the PA that the flight is delayed for another 30 minutes so you open up your presentation and make a few simple changes until you are happy with the final result.

Since you no longer need to make any changes you decide that you will re-upload the file to the cloud storage (Mistake number four) and then grab yourself a coffee. You do and sit back to enjoy it and by the time you have finished, your plane starts to board. You pick up your carry-on luggage and board the plane to Melbourne. None the wiser as to what is happening while you were in the air, you arrive at Melbourne and grab a taxi to the office but when you arrive at the office something strange occurs. The cab fare is $43 dollars and when you go to pay with your EFTPOS card it declines. That can't be right you checked before you boarded the plane and there was almost $2K in your account. You shrug it off and pay with your credit card and make a mental note that you will have to check what happened later after you have your first meeting.

You arrive just a few minutes before your potential customers do and get ready for them in the boardroom. You connect your laptop up to the network and open the presentation (mistake number five), do a quick run through to make sure you are still happy with the changes you made to it before you boarded the plane in Brisbane which you are. a few minutes go by and then the board room phone rings and its Janet from reception telling you that your guests have arrived. You go and meet them in reception and then head back into the boardroom with them to start the pitch. Everything is going well until about 30 minutes into the discussions when suddenly your laptop starts to freeze and behave badly. Then an image comes up on your screen that says "your system and files have now all be encrypted and you need to reach out to ***** email address to negotiate the cost for decryption of your files" (or something along those lines).

The same message starts to appear across all the machines in the business at the Melbourne office and then a few minutes later starts to spread across the private VPN between sites and all machines across all sites have the same message. All files are encrypted with no access. As you would imagine the meeting was over and it's probably likely that they won't be doing further business with your organisation.

So, you can see that this issue escalated quickly but did you know why I was indicating on each occasion why the businessman – let's call him Harold made a mistake? Let's look at each occasion and then discuss what happened that Harold didn't know at each point.

Mistake number one – connecting to the free wifi. At this time what Harold connected too, it was actually a connection set up by a malicious actor, so that they could collect and capture any traffic that anyone using the network was transmitting. So that then brings us to mistake number two – Harold then logs into the corporate network giving the malicious actor login credentials to the cloud platform.

Mistake number three - Harold logged into his personal bank account giving the malicious actor access to the account. If you remember when Harold tried to pay his card declined because why Harold was in the air flying from Brisbane to Melbourne the malicious actor transferred all of the money out into an offshore account and then probably moved it another ten times before finally exchanging it for bitcoins which was also then transferred several more times between different bitcoin wallets (you could say that money is gone forever now).

Mistake number four – Harold uploaded his updated file to the company cloud storage which the malicious actor intercepted and modified to include a little something extra (Ransomware bug) before sending it onward to the cloud storage using the stolen credentials that Harold had already given him (remember mistake number two).

Mistake number five – Harold connected his machine to the corporate network and then executed the modified version of his presentation on his machine, thus executing the virus and as Harold continued on his normal work the ransomware bug has started to do its worst in the background and well you know the final result. Everything was encrypted, and it was all possible because of that first step. Connecting to the free wifi connection at the airport.

This is obviously a worst-case scenario, but you can see how easy it was carried out and the malicious actor didn't really have to do much at all to make this happen, Harold basically gave them keys to the network and said go for it without even knowing that he had done it.

You are probably thinking okay great I get it, the house burnt down, and it was all Harold's fault and technically yes it was Harold's fault but that is not the lesson here. Harold should have received awareness training from his organisation and he should be made aware that he should NEVER connect to free wifi. Harold carries a company smartphone that has the ability to be used as a wifi hotspot and share access to the internet. This is what should be done at a minimum. This will stop the scenario at mistake one and prevent that day which could have been the best day of Harold's career.

Obviously, there are some issues with the way the company has set up the network that allowed it to spread right through the organisation and a lack of good quality antivirus/IDS/IPS that could have stopped or at least minimise the effect, but I want to leave this at the free wifi, for this is the lesson I want you all to learn. That alone could save you from a similar fate as Harold my poor imaginary businessman.

I want to describe another scenario for you now just to ensure you really understand the dangers I am trying to bring to your attention if you use the free wifi. Let's look at a motel, they could have hundreds of guests stay over a week and guests expect to have fast internet available to them when staying but should you use the free wifi? NO, never use the free. Let's look at the motel free wifi for a moment, if you have 30 guests all connected to the wifi at one time (it will probably be horribly slow but that isn't the issue here), as a malicious actor I could do the same scenario I described in the airport and just capture all data on the network.

I could also scan the network and gain access directly to machines on the network to steal data or infect them to spread my viruses or expand my access even further. I could go on for probably another ten minutes on ways that this could be used to my advantage but by now you must have started to understand what I am trying to get across to you all by now.

Never ever use free/public wifi connections it's not worth the risk, use your mobile as a hotspot, buy a mobile connection that can be used outside of the office. Just remember the ease at which the incident could escalate and do the right thing here.

Chapter 41 - Updates are like vaccinations

Updates are a funny thing, we all know that updates are released for our systems to fix bugs, security issues or add improvements to the functionality. Basically, it makes updates like vaccinations, right? Vaccinations inoculate us from getting bugs or at least minimising the effect they have on us and that is really what updates are about. So why don't people do more updates?

We are told time and time again that we need to get the basics right in order to ensure that our corporate systems are as secure as they can be, but it just doesn't happen. Look let's be completely clear here, most attacks that are still physical attacks on systems (not counting social engineering attacks) are problems and vulnerabilities that have been around for quite some time and already have patches released that basically make your systems immune to the attack or specific threat.

Honestly there is millions of vulnerabilities out there that could be used to crush your network in minutes, absolute wipe out and it would be too late for your teams to do anything to stop it but if you had just taken the blue pill (matrix reference – I am a bit of a fan) and vaccinated your system none of it would have happened.

When it comes to security everyone wants to throw money at all of the new blah blah blah with all of the blinky lights that supposedly is better than all of the previous 50 versions of the same thing with a new name and pretty GUI for you to look at. Yes, disclaimer here some of the blinky lights stuff that all of the vendors want to sell you are getting better but throwing stupid amounts of money at the biggest and best security solution that is being touted as the saviour of all our security concerns will do probably two things in most cases – lower the bank balance and probably disappoint you when it really needs to do what it broadcasts to the world that it can do.

It is honestly a sad truth but if you haven't actually spent time working on your security basics like asset management, system updates, passwords policy and system hardening just to name a few things, you will fail no matter how much money you have spent on these new fancy toys. Honestly, if you can truly ensure that you get the small and boring stuff right you will probably have a stronger more resilient system than someone who has skipped all of the prep work and just moved straight to the flashy lights. I am being completely serious here, all the money in the world won't save you from being lazy and ignoring what you should have done in the first place.

I know some of you will whine and complain about the fact that you have legacy systems that can't be updated for some reason or another but that is a cop-out. If the systems can't be updated find a better solution or isolate it so that nothing but the bare minimum required can even know it exists. I think that if you truly want to work out how to get your systems up to date then you can do it. Yeah, it is not going to be easy and it may cost some money to get this done right but you will be much better off for it that I can promise you.

So, if you don't have an excuse for not doing updates and they will basically make you the most secure you could be without throwing a ridiculous amount of money at the problem, then why do some many companies fail monstrously at keeping systems updated. You can automate most of the process with asset management systems so that once it is set up correctly you will barely even need to do anything to keep them continuously updated.

Do yourself and your organisations a favour stop and plan what you need to get done to get all the boring foundational security stuff sorted then by all means throw some money at some awesome new blinky lights to improve on that foundational work you just got through getting right. Look I love playing with their new toys too and seeing what can be done to put them through their paces, but they can't replace true security foundation work if you remember that you will be ready for that next wave of attack and with a bit of luck and your hard work will have all been worth it.

Chapter 42 - How can we help our society's elders know their risks?

This topic is very important to me and I really feel that we need to evaluate what is happening and find a way to make it stop. I hear quite a lot that elderly people are being ripped off via scammers with some sort of social engineering attacks. They will in many cases call up and say they are from Microsoft or Telstra or whoever is the latest company they want to pretend to be from and they convince the elderly person that they need to give them access to their computer or get them to give them account information.

These people whoever they are true scum, how on earth can they sleep at night knowing that they ripped off probably between 10-20 elderly members of society who have worked hard and done their part for our society who may not afford to eat now because the criminal thought they needed to have what these poor elderly people have more than they needed to keep warm.

I just get really angry thinking about this so a warning this topic may be a little more aggressive than usual, but I will try to keep it civil. So, this scum of the earth gets them to give them access to bank details or gets access to their computer to either get that same information or use that machine for whatever other purposes they are wanting to utilise it for.

Could be ransomware, crypto mining or so much more. Whatever the end result it is a real hindrance to the victim and we need to find a way to educate these elderly members of our society that they should forget the belief that they can trust people. Nothing in our society is as it was when most of them grew up. Back then you could walk out of your house with the doors unlocked or even open and you would not have any risk of someone walking in to steal your television or whatever else they would steal back then. That type of behaviour was not something that happened, people didn't do it.

Back then if someone came to your door or called you on the telephone and said they were from the bank or the phone company they were actually from the company they said they were, there was no social engineering attack on our society that is happening as we speak. So when you try to tell them that you shouldn't believe them and you should hang up on these scammers, it's just against their instincts. It is probably completely alien to them and I get it.

We can't leave it at that though as they are suffering, and the number of victims is constantly growing, so what can we do to stop it. I am not sure what will work but I know that we need to really come together and find a way to make it work.

We could look at running an advertising campaign on a medium that is popular with the elderly target market or send out flyers and promotional material to them that is created in a way that will hit home with that group as nothing that I see has made any real benefit. Maybe we could look at putting together free training sessions on these types of scams and provide the service to all senior community groups to help spread the message as much as we can. Maybe these will work maybe they won't.

Maybe we are looking at this all wrong and are focussing on the wrong group to get education out to these groups. Should we instead aim at pushing education towards millennials and get them to help educate their parents and grandparents on best practices and how they should shut down those scam conversations before they become a victim?

None of this is a guaranteed solution to the issue but if we do nothing we will be certain to achieve nothing so its better to put in the effort and try to make a difference and rid this scum who is looking to rip off your parents and grandparents of all their savings (that's probably your inheritance gone if they get scammed – so there is some self-motivation if you need it).

Let's come together and put something in place. If it fails to make a difference lets just, try something else. Nothing ventured nothing gained I am told so let's do something about this today.

Chapter 43 - Social Engineering Attacks

Over the years I have heard of some really well-crafted social engineering attacks, it is a really great way for malicious actors and Pentesters alike to get past well-trodden security measures. To be honest there really isn't too many proven systems that can defend against talented social engineering malicious actors. They know what makes humans tick and they will abuse every possible emotion or human nature angle they can twist to their benefit. Talented ones are amazing to watch and can be a little terrifying to see how easy it is for them to worm their way past checks and measure with ease.

One that comes to memory that was a pretty amazing and personally the guy who put her up to the attack was shocked you could see it on his face. I really don't think he thought she would be that successful, but she nailed it. Jessica Clark the social engineer who completed the attack pretended to be his wife and was using social engineering at his request to try to gain control of his accounts – I bet he is feeling stupid about that now. She puts audio of a crying baby on in the background and in less than 3 minutes has complete control of his accounts. It's funny, check out the video on YouTube.

Another one was on a CNN report in which a team of professional social engineers demonstrated how easy it was for them to spoof their phone number to appear as though they were calling from a number within the target organisation, they then called through to the IT support desk and started saying that they were not very good with IT and they can't get access to their project they have online. He then continued to give them a web address and got them to click to open the file. The tech just gave him access to his machine and a fake project file come up and the guy said that he must have done something as it works now. The tech on the other end said no problem and was none the wiser that he just let the attacker into his systems. Check it out on YouTube this one was even faster with no tricks, just simple manipulation and some number spoofing.

People put up big expensive security perimeters around their worlds thinking that they are impenetrable but that is far from reality, the above examples are just two of hundreds or even thousands you can find on YouTube or social media (Don't believe me – do a quick search). So, we need to ensure that we have a plan for social engineering, set up a plan for how we can best manage the processes to ensure that systems are protected as they can be.

I am not talking about more blinky lights and more security firewalls or some other fancy AI, that won't protect you from social engineering (okay so there might be some technology that may help but that's not what we are here to talk about). You need to train your staff and train your staff until they are comfortable with detecting these types of threats and then train them some more just to make sure. Security awareness training is great, but this is not about meeting compliance which most companies just use it for, this is supposed to help your staff be better at identifying risks. Help them to better understand how their actions can put the company at risk and what can be done better as a team to improve.

If people don't understand help them, don't belittle victims, use an incident as a learning tool so they can know what happened and how to spot the issues next time. Spotting social engineering is a learned skill, even us security folk had to learn it at some point, how else would we know how it happens? So now we have done all the training and our staff are aces at detecting social engineering attacks mistakes will still happen. So, train and train make it fun but don't stop at that.

Review your daily processes that could be manipulated by malicious actors, run through them and outline the particular stages that could be a problem. Then change the procedures to include secondary or even third or fourth checks in high-risk cases to ensure that staff can't be manipulated easily. Ensure that these new processes are implemented and have them tested if they don't work and can still be manipulated somehow change them again and again until you have a solid process that will reduce the risks of these attacks being successful.

In these procedures ensure that all financial records that there are always two staff members involved with the process and ensure that customer or supplier checks are in place if requests are sent through specifically about payment details changes. It will ensure that your organisation and your suppliers/clients will be better protected by your processes.

Now once this is done and you have the training flowing nicely, the procedures are practised and hardened then it can be a time to come back to looking at the blinky light solutions that could extend the protection for your staff. You can get some great email filtering services that can help to reduce the number of malicious social engineering emails that actually reach the targets email inbox (I am not going to name them as I am not here to sell you solutions that is not my goal at all, I want to help you be better and protect your organisations). They can include click protection and spam filtering and some are now also including a training section to teach your staff as they go while protecting them from incoming threats.

I know some of you are probably moaning and groaning right now saying that all of this is too much hard work but I am sure unless you have been hiding under a rock lately that you would be aware that there is at least one high profile breach every week and they are just the ones that make the news, so I am sure you can imagine that the number would be dramatically higher as the news isn't concerned by Tim's pies or Jill's flowers but we need to be.

So, quit whining and make a plan to secure yourselves from social engineering attacks you will be much better off in the end. This isn't just about your business though as if you train your team they will pass on their new skills to their circles, in turn, making everyone that little better prepared.

Chapter 44 - The Life of a CISO

A CISO (Chief Information Security Officer) is the pinnacle of successes and the aim for many a security engineer in the industry but many will never achieve this level of success. However, is this really a problem? I am not sure it is, the CISO in today's tumultuous security environment, in my opinion, is probably one of the most stressful jobs you could have so why would anyone want to put themselves into that situation on purpose? It seems madness, doesn't it?

Let's look at this deeper though and really see what life could be like for a CISO. Let's look at Jim, he is a CISO for a made-up company called Software house. Software house is a company that developed a medical practice software that is used by thousands of doctors surgeries around Australia (very sensitive and critical information there I would think). It's a very successful business and is expanding its market share on a daily basis. Today however, Jim's day starts at 3 am with a phone call from his on-duty security staff. The company's systems are in meltdown, nothing is working, and they suspect that it may be a security breach. He gets out of bed and got ready quickly before he makes his way into the office.

On his way, he notifies the rest of the security incident response team and gets them on their way to the office. Today is really not going to be a good day for the security team and maybe not the organisation as a whole. Jim arrives to a chaotic situation in the office with what now looks to be a crypto virus that has spread through all of the systems on the network. The security team was unable to stop the spread of the infection and every machine on the network is displaying a "you've been hacked" message on the screen. We can be certain that this is not a good sign.

The company has no current backups, as there is no network isolation and the encryption has reached all of the backup storage (That's an oh crap moment if I have ever seen one). There are no external offsite backups due to budget restrictions (the organisation didn't see the benefit and preferred to put the budget into marketing and development). To make it worse all client data is hosted on the organisation's servers and clients don't have local copies of the data as this is a service the organisation sold as no maintenance required, The Software House will handle all that for them (That was a big mistake).

Jim has recommended changes every month at the board meetings but every time the increased budget request to ensure separation and adequate data protection measures were denied. They may have been the saving grace right now if they had been implemented. Jim and his team locate an annual systems backup that was stored in the company safe, but it is now August, so it is 8 months old. That is certainly going to be detrimental to the organisation if they need to restore to that snapshot, but it is looking more and more like it will be the best option.

Jim advises his team to reach out to the malicious actor and obtain a price to get the files unlocked while he goes and wakes up the board to inform them of the situation, this is not going to go well especially at 4 am. As expected it went delightful and I am sure that Jim is not on the Christmas list this year for any of them. Around 5 am the malicious actors come back with a request for $100K to unlock the data which is to be paid in bitcoin. The board has arrived in the office and Jim takes the request to them to decide if they want to pay the ransom or not (Jim strongly suggests they do not pay it – but they decide to pay the money anyway).

The payment is made in bitcoin to the malicious actor as requested but days have gone by now with the systems still being offline and no response from the malicious actors at all. Dead silence once the money was received. It would appear that they have just taken the money and no data will be released. After further discussion, the board decided to restore the data from 8 months ago, but it is already too late the business. Systems may be back up, but 8 months of data is lost for all of their clients. That's is an epic disaster and that will not go well with clients who have been calling non-stop since the incident occurred. Silence has been the direction of the board against the advice from Jim so when the incident is announced it is sure to be an onslaught.

The Software house will un-doubt ably throw Jim under the bus for this incident as they have already asked for his resignation. No blame is going to flow uphill from this, the board is going to make sure of that but who is really to blame in this situation. The CISO has made constant failed requests to make improvements to ensure that systems are recoverable if an incident occurred, but they were always denied by the board (is that the CISO's fault for not pushing the urgency or the board for not seeing the need to invest in cyber protections). When the board was advised that they should not pay the ransom they still proceeded to do so and then they asked for the CISO's (Jim's) resignation when they didn't unlock the data (I am not sure that is very fair).

The board needed a scapegoat for the storm that is now going to overwhelm the software house and Jim is that man. So, after that nice little story, do you all still want to be a CISO? Yeah, you probably do and so do I. I act as a virtual CISO in my day job and although it can be painful and a mountain of stress sometimes it feels good to make a difference in this cyber battle we are all waging against the malicious actors.

Hopefully, now you can see what the life of a CISO is really like. 16 hour days, unlimited stress, probably be denied all of your security upgrade requests but when all the S*!T hits the fan you will be the one that they will throw under the proverbial bus. Okay, this may be a little dramatic, but this scenario is becoming more and more common every day and that needs to stop.

Do the right thing with regards to your organisation's security team, not just when things are good but when they are really heading down the rabbit hole, look after each other and don't throw your team under the bus. Stand tall together knowing you have done all you can. If you are the CISO it doesn't mean that you are to blame for all the mess (Yes, you might be but probably not). So work together to make your businesses safe and if your dream of being the CISO someday at least go into it with eyes open and truly understand what it is you may be up for.

I hope I haven't scared you off with this one, oh well.

Chapter 45 - Passwordless Future - Is it really an option?

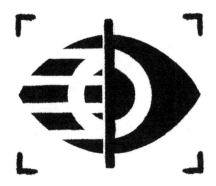

Passwords are an archaic and painful way for users to securely authenticate themselves. They are one of the biggest issues that security teams need to deal with in an enterprise, they are constantly stolen or cracked due to user poor security practices. You know the things that I am talking about, using the same passwords for all your accounts or just changing a number by counting up and down at the end or the start of the password.

What about those of you who use your pet's name or date of birth of one of your children or your partners (don't do that it is insecure), then we have the one who uses a variation of password or qwerty (I am just not going to say anything more about these). You all know someone who does one or more of the above password options and you probably know because they have told you what the password is (I just literally slapped my palm against my forehead – wow this is as bad as post-it notes on screens which sadly happens more times than you would believe). So, I think you get what I am saying here, people really suck at good password security.

This is partly our industry's fault and I will explain why just to give it some more context. We in the industry have been telling everyone for years and years to create a password that has numbers and characters and symbols so that it is really difficult to break but there is a huge flaw in this that we all just completely missed (honestly, I just followed the norm and recommended best practices – which is my own fault for not thinking for myself and just being a sheep) these types of passwords are really hard for us humans to remember and then we want you to change it every 30-90 days just to be a bit more of a pain in your sides (than we wonder why people always need password resets).

These types of passwords are not exactly the most secure option either, with the power of today's computers they can be cracked in a reasonably short amount of time. Let's do some quick numbers on this, I have a password – Secure134! It has a substitution and some numbers, and it would be what we would call in today's best practices a pretty good password. With a medium-sized botnet (nothing fancy but something every decent malicious actor should have) they could breach this password in 6 days, parallel GPU's could do it in 87 years (that is not very fast), so I would think that a rig set up with 10-12 aligned GPU's it could now be done in a few weeks.

I know that sounds like a long time but for a high-value target, that really isn't much of an effort for a malicious actor when all you need to do is set it up and leave it to run. Chances are it could be done in under the password policy expiry time frames. Why are we stuck with passwords like this when you could actually use a passphrase, it is much more secure and can be much easier to remember. Let's do an example of what I mean, take four random words or more and string them together like – blue mustang river steaks, this password with a medium sized botnet would take 52 billion years to crack, 260 trillion years with parallel GPU's. Now that is clearly a much more secure option, wouldn't you agree?

Passphrases are a great option but, in the end, they are still just passwords and are in turn still prone to be lost or stolen or you name it. So why don't we just stop using them? Well, it's because people don't like change and it's the easiest way for systems to provide secure access without too much effort or costs. With the obvious flaws is because it is easy for a good reason to keep using them? No, it probably isn't, so if we decide to change and stop using passwords all together can we really do it?

It is certainly an option that has been thrown around a few times but passwordless access controls are certainly a future option that has never truly been in our reach but more recently it has become something that is almost achievable. There are some USB key fobs with fingerprint scanners or we have facial recognition options and fingerprint/vein pattern scanners which are okay. Combine a couple of options together with biometric and maybe a security authentication app then we have a workable solution but is good enough?

Is good enough though better than current password practices, no I don't think it is. I personally still feel that a combination of a good passphrase and a second or maybe third-factor authentication method is still a more viable option to individuals or organisations.

I honestly don't think the technology is quite there yet and believe that in a couple more years passwords will, in fact, become a thing of the past (that will make it a bit harder to be a malicious actor, but I don't think I will be shedding any tears for them). This move away from passwords will be a good move for security however, it will not be the end of security breaches as malicious actors will continue to find new methods of attack and passwords are just one method in today's systems.

They will undoubtedly still be riding the social engineering wave that's for sure and I can almost guarantee that there will still be software vulnerabilities for them to exploit as there will be certainly organisations who won't be doing the basics with regular security patching. I don't think that will ever change no matter how much we say it's needed, we will go blue in the face and collapse before we will get that message across but that doesn't mean we shouldn't keep trying.

Let's keep our eyes on the passwordless options and when the time is right to make that move but until then please can you all try to be good corporate citizens or cyber individuals and actually practice good security practices.

Chapter 46 - Smart homes - Your security death trap

This is not the first time I have written about smart technologies and the risks that they can pose but I wanted to narrow the view a little with this particular piece and focus specifically on smart home technology and some risks they pose to you and your families that you may not have considered but really should. Don't get me wrong I am as intrigued by smart technologies as probably many of you are and what benefits they can bring to our lives, but I just don't know if the benefits outweigh the risks.

Let's break these smart home technology devices/functions down into a couple of different areas and then review each for the possible risks you should know about. The first is the smart locks, home security monitoring systems, then we have systems like the smart fridges, freezers or random cooking devices or coffee machines (some of these are just loopy but they exist). Lastly, we can then look at smart speakers, home hubs, child monitoring devices (the risks with these really concern me – so I might start with this one first).

Smart speakers are something that is spreading into homes almost as fast as the first iPhones spread into our hands when they first appeared in our lives back in June 2007, I wasn't a big fan when they came out, but they have certainly brought about a big change in how we all use our phones. Smart speakers are actually a pretty cool device and are quite impressed at how well they can actually follow voice commands and answer questions thrown at them but is it worth playing some music via voice or answering trivia questions to allow them to record all your conversations that are in microphone shot of the devices? Then for the companies who make them review and use the information for their own benefit with absolutely no regard for our privacy.

Just one example of this was with Amazon Echos, with more than 100 million of them sold by the start of this year (2019) worldwide, wow that is a massive number. It was revealed that Amazon employs thousands of staff around the world to listen to recordings captured unknowingly by echoes of its owners at homes and offices. That certainly sounds a little stalkerish, unethical to me and a massive breach of consumer trust by

Amazon. Amazon says its to help the AI function better respond to commands, but I think there is probably more to it than what they are telling us, but it will be forgotten, and nothing will be said about it again until the next time one of these massive companies do something like this again (Oh and did I mention that they haven't stopped doing this either, this is still an ongoing practice).

Smart baby or child monitors now this is truly an area that concerns me, this is our children that are watched and listened to by these devices. These devices are put in our children's rooms by parents and then connected to wifi, so they can access them via an app on their smartphones through the internet and interact remotely with their children, no security setup on many of the devices and parents installing them on many occasions don't even change the default login information. This is a huge risk to your security, personal privacy and possibly your children's lives if these devices are accessed via a malicious actor. Don't believe me take a look at this case with the Nest baby monitor a malicious actor threatened a mom that they would kidnap their baby, that would be a bit concerning to a parent don't you think.

It doesn't end with baby monitors though what about smart toys like those reactive toy dinosaurs or stuffed toys that we all see in the toy stores these days that interact with our children. They are connected to wifi in many cases (they say it is for updates, but I don't think that is the end of it for some) and they can be accessed via an app on a parents or child's devices to control or customise these devices. They can also be used by malicious actors to listen in on children or their families and who knows what else. Please, if you must use these devices in your homes at least change the default login information and if you don't know how to do it please ask someone who can help. Seriously this is our children's safety you are talking about here, don't just fob it off and say it will be fine. Trust me it won't be fine, secure these devices, please.

Now I want to be clear here I have barely scratched the surface of what is capable by malicious actors with these children/baby devices but if they are capable of activity such as above with let's be honest here pretty dumb devices what do you think they could do with your smart speakers or let's say you're smart fridge? I did a previous topic of "Hacked via a fridge" I think that clearly describes how bad it could really get with those types of appliances. I personally like the fact that not only could malicious actors use your device to hack the network, but they could be super nice and order you 500 bottles of milk or 100 boxes of Magnum ice creams just to make sure you don't run out. That's pretty nice of them don't you think.

Do you really know what information your coffee machine collects about you and what it sends back to the manufacturer? Honestly, do we really need to make our lives more insecure by making smart kettles or toasters, come on people we don't need that type of connectivity in dumb devices like these, it's just idiotic! What could be a benefit, it could tell me I haven't got enough water or maybe I put a bit too much in? Just use your eyes and your brains it's what they are for.

Okay, so you have played around with the smart speakers, smart tv's and you have a fridge that orders your groceries for you, so you thought let's go the full hog and fit out our house with smart locks, lights, garage doors and security monitoring systems (I can only see this ending badly). So, I get people love to have things automated and have it so that if you want to check that your garage doors are closed when you get to work when you just have that nagging feeling that you left them open. The ability to remotely close them or open them if you get a delivery and you discuss the placement of these with the delivery person via your connected doorbell. Yeah, I get it, it's a cool feature and could come in handy.

Maybe it would be great to be able to view your security cameras via your phone and turn off the alarm as well as unlock the door to allow a tradie to come in and fix your washing machine or toilet or whatever it is that has notified you via the smart function that it has a malfunction but stop here for a moment and think about what I just said. Just so we are on the same page here I am going to just create an imaginary scenario for you.

Let's say you have a smart home that is connected to basically all of the above and you post on social media that you are having a super awesome holiday in the Canadian alps (seems like as good a place as any) and a malicious actor who has been watching your profile for a few weeks now and already has broken into all of your accounts online (it wasn't hard you use the same password for them all including the social media which was involved in a breach last year – you should have changed it by now and not used the same one for everything). They already know your address and as they have access to all of your accounts they rock up to your house in the middle of the day and turn your security systems off (so you can't see them on recordings), then they unlock your smart locks and walk right in and help themselves to all of your fancy new smart devices.

All these fancy smart devices in your home don't quite seem as smart now do they? No, they don't. That is just one type of scenario that these devices could be used against you, a stalker type situation could use these devices to watch what you and your family do at all times, track your locations with smartwatches or phones. The list really does just go on and on. My mind runs wild with scenarios in which I could take advantage of these devices to gain access to systems, I am not supposed to, and I am one of the good guys, so I am sure you could imagine what sort of scenarios that a malicious actor could come up with such a plethora of unsecured devices to manipulate.

Okay so I may be being a bit dramatic, but these threats are real, and you need to ensure that you take all the necessary measures to protect yourself and your family if you don't know how please find someone to help you do this. Even if it costs you some money to ensure you are secure it will be definitely worth the investment. You can't put a price on our children's safety, so go do something about it today and stop putting it off.

Chapter 47 - The Dark side of the net

The dark side of the net or dark web as it is also known is a mythical place in which demons, fairies and unicorns all mingle for late night karaoke and fosse ball tournaments. A place of magic that is seen by a meagre few who live to talk about it. It is like the Bermuda triangle of the web in which many who venture in just seem to disappear in the abyss, never to be heard of again. The Dark web as a place is shrouded by so much hype it is difficult for anyone to know what is real and what is just a mythical fairy tales that were invented to scare children and non-hacker folk at night so that they would never dare to attempt to enter.

I find it amusing when I get people asking me about the dark web or darknet and the dark mysterious shadow it seems to hold in most people's minds. It really isn't a mysterious magical or scary place that is too often made out to be. Honestly, it really isn't.

PUBLIC SAFTEY WARNING NOW (or personal safety at least) - Okay, so before you all throw caution to the wind and dive into the dark web, this is still a place that you should be wary of and use some caution before doing anything stupid. I will explain so you understand why I say this after just saying that it is not some mythical horror setting that it has been made out to be.

The darknet is not all bad but you will still find many a dark market place in which you can buy anything from drugs, weapons or stolen passwords or hacking tools. Almost anything you could think of, you could probably find someone that you could buy it from. Although there is much more than just illegal market places on the darknet and is a fast changing playground for anyone who wishes to keep a feeling of privacy in a world in which privacy is almost completely eroded. Anonymity is the whole reason the darknet exists, as it is known as today.

A big percentage of users are from countries that are not so friendly towards free speech and really like to control what their citizens are allowed to access. The darknet is perfect for these people as it allows them to access almost all of the same content that could be accessed on the normal internet experience but minus the chance of imprisonment or lashings or worse if they are caught accessing any unapproved or non-filtered information. It is sad that this is needed in some places around the world but not all countries are as easy going as Australia.

It is actually quite simple to access; you just need an onion browser and a good VPN. Trust me even though it will work without a VPN you don't want to put yourself at risk by not hiding your real IP while looking around – you don't want to put a target on your back with a hacker or worse by saying or doing something you probably shouldn't have.

The hardest part is actually finding content, the darknet is not crawled and indexed by the search engines like the surface web, search engines like Google or Bing basically ignore it. You can find some directories that can help you navigate to popular locations, but you will quickly be frustrated with the fact that many sites will constantly change site addressing, some to help keep their dark marketplaces hidden from authorities others just to try and protect themselves from DDoS attacks as these are a huge problem in these dark places of the web.

You may have an awesome site that you found today which has all the information or products you want that will disappear tomorrow with no notice or warning and you will probably not be able to find it for months or maybe ever again. It can certainly be frustrating to say the least but if you can deal with the clunky annoying fact that everything will take you three times longer to do it can be a good option.

So, get yourself an onion browser and cautiously look around, it really is not that scary but if you are going to look at criminal marketplaces, you need to do it with your eyes open. This is not like buying things on Amazon or eBay. I think the main point you need to be aware of is that anything that you read or see on the dark web/net is probably a **LIE**. Do not trust anything at face value, you could have sellers that could have perfect five-star ratings but they will quickly take your money and disappear as much as they will deliver whatever it is that you have come to this dingy place in order to find.

You really need to do your homework and be cautious if you are going to buy anything or even sell items in these places otherwise you will be quickly stung by a really bad experience and may regret your decision to venture into the dark crevices of the web before you scurry back to the standard web quicker than you can say "I will never buy anything on the dark web again".

If you can get your head around this fact and understand your risks, you really can find anything you want, just take my advice and do not do anything stupid that will get you arrested or something worse as it really is not worth it to buy something that you really shouldn't.

So, brush away the silly myths that you may have heard and have an adventure into an unknown cyber realm if you dare you might be surprised what you will find.

Chapter 48 - The Race to become more than human

204 - Craig Ford

Machines have come a long way over the last 20 years and it would be naive to think they will not continue to advance with the same leaps and bounds in the next 20 years. I think some of these advancements will be amazing and will allow us to really achieve great things, but one area is a little strange to me. The push to merge humans and machines is becoming a highly discussed topic. I regularly see ideas of how this integration could develop and it will only be a matter of time until the line between human and machine is almost unrecognisable.

Exoskeletons are something that has been a regular in many a Hollywood blockbuster (one with Matt Damon in from a few years ago rings a bell - Elysium) if you have seen the movie or not doesn't matter but I personally liked it. It had some pretty awesome fight scenes in it, Matt Damon can do the action stuff that is for sure. However, the extent of integration that this exoskeleton had with Matt's character was very impressive, to say the least, and although the movie was obviously science fiction these types of suits are starting to be tested and integrated into our lives more and more.

We have systems in use today that are allowing paraplegics to rise up out of their wheelchairs and walk again. Yes they are still very clunky and not even close to the normal function of movement that the wearer would have had before they lost the use of their legs but it can be certain that these exoskeletons are moving forward quickly and are making massive strides forward so it is certain that it won't be long until users will be able to function as well as they did prior to the injuries with these suits (Honestly though they may not need to if they can find a way to fix this medically). However, even if medically these suits or exoskeletons are not needed the advancements in soft exosuits that are designed with artificial muscles and ability to reduce the wearer's fatigue and strain.

Military (you had to know they were already using these, right?) they are fitting out soldiers with exosuits that can enable them to run with massive loads of equipment for hours with very minimal fatigue. This means they can be better protected and have all of the equipment they may need to handle any situation, no compromise and no fatigue problems. Sounds like a win-win to me.

Medical lift aids are allowing nurses or even patients themselves into and out of bed, showers giving them easy mobility for situations that may have previously required two people to manually help them in these situations. These situations would have also made the carers prone to injuries themselves and if they are hurt how can they help someone they are caring for into a shower or into a wheelchair they wouldn't be able to. These exoskeletons or exosuits make the wearer as strong as 3-5 men with the ability to lift, so they can pick up the patient as though they were lifting up say a small child. No injuries, no strain just simple and effective help with minimal risks to all involved. Perfect scenario if you ask me.

I believe all of these scenarios are great and they can benefit many areas of our lives but currently, battery storage is a definite problem that will need to be resolved before they will be able to become extremely effective in many fields. Possibly even some improvements in power consumption could allow them to be more integrated into more industries that require a high level of physical labour.

Even Elon Musk is getting into the game for further integrations between humans and machines. Elon is working towards a type of neural lace concept, now this is a pretty complicated idea he has and I will try not to butcher it too much here but in its simplest form Elon wants to connect our brains directly to quantum computers to basically eradicate the roadblock that our bodies put on us, basically when we communicate to computers at the fastest speed we can type or maybe talk. This is so slow in comparison to the processing power of our brains or some of the most advanced computer systems today so why not cut the roadblock out and directly communicate between our brains and computers we wanted to do our desired tasks or compute our most extravagant calculations.

If this type of direct link could be created, it could drive technology forward drastically. Seriously the possibilities are just mind-boggling and honestly, Elon is one of those people that could probably bring it all together if he can juggle all of the projects he is running with at this time with autonomous cars, space flight, solar and batteries. I hope that he can

206 - Craig Ford

pull it all together and succeed but we need to see more than the awesome tech here, what about the risks? If humans and machines get this direct link and we have a get real AI what is to stop the machine taking control of the human? Or image the damage malicious actors could inflict on the world when they can act as fast as they think.

That whole thought scares me, some of the malicious actors out there are really smart people that when provided almost a direct channel to the systems of the world could do scary and amazing things that no matter what we do could not be stopped. I think security is tough in today's technology era I can only imagine what it will be like when you really can't see what part is human and what is the machine.

One concern with Elon's neural lace concept is that currently, the plan involves the user to have surgery to implant chips in their brains. That's a little ridiculous and a bit scary if you ask me, they need to come up with a much less invasive option for that communication channel don't you think.

Neural links and exosuits are not the only blends between man and machine. People are embedding chips under their skin to make payments or public transport easier (you will not get me doing that) then we have athletic suits that are like a second skin that keeps us cool or warm when our body needs it, monitors our vitals. I even believe one option had a defibrillator built in case of cardiac arrest.

We have robotic arms and legs that have direct brain links that users can control like our limbs we were born with, they are even making them capable of letting the user feel touch. They are really doing great work in these areas but is it sad that every time I see something like this, I automatically think about how I could hack them, what security protections do they have in place if any at all?

Don't get me wrong here I think all of these advancements have great potential and can help make life easier, but we need to ensure that we do them right in the first instance, protect them properly, don't take shortcuts or it will only take one opportunity for a malicious actor to take advantage of a weakness. (As I wrote that line I was smiling as I had an image in my head of remotely taking over an exosuit and forcing the wearer to do the chicken dance or something).

So, let's ensure that we push the boundaries of what is capable with the merger of humans and machines but don't let it be at the cost of security. I am certain that we can do both without too much hassle (oh and I am still having a giggle to myself about the chicken dance idea).

Now the next two chapters are a little different and are what I would call rough ideas for two hacker fantasy books. They are a taste of a possible hacker fantasy series that I am starting to put together. I hope you enjoy the small taste of what is to come.

Chapter 49 - Journey to the dark side - a hacker fantasy tale

Tick, tick, tick the silence is almost deafening. I have been sitting here for hours and I cannot figure out how I got here. Not to this place but in this situation, I just don't understand where it all went wrong, what set me down this path. Was it one choice or was it a set of decisions that made this path a reality, could I have changed my fate or was this always where I was meant to be? The metal around my wrist is starting to really dig in, the fed that put them on me was really young, to be honest, I don't think he looked like he was old enough to have finished high school let alone be an AFP officer. I guess that means I am starting to get old since almost all of the feds in my house seem like they are a little older than middle graders, it pretty much cements that idea. Yep, I am the old guy in the room and I am not even forty yet.

I think he thought I was going to try and escape or it was the first time he had ever put cuffs on a real suspect before, if it was indeed his first time he has done a pretty good job. These things are going nowhere, and I guess that means neither am I unless this bunch decides that they have made a massive mistake and just let me go! Hmm, that will certainly not happen, well at least not yet anyway, they think they have me. I can see them collecting up all of the electronic devices that are in my house and tagging everything. They are certainly ensuring that chain of custody is maintained, I guess if they truly do have me, which they don't but if they really do then I am definitely not getting off on a technicality that's for sure.

If only they knew that this is a complete waste of their time, they won't find a single trace of evidence on any of these devices. If they are lucky they might find some traces of some porn or torrents from my housemates, honestly those boys get up to some funky stuff sometimes. I can only imagine what sort of stuff they get up to on the net, these AFP stooges are really going to have some fun dredging through all of that history. That isn't what they are looking for though, they think I am the hacker responsible for a string of breaches on several Australian politicians, some really funny images of them had been put up on their social sites and they made some very unprofessional posts online with them in some very compromising positions. Look even politicians have sex but those thugs were stealing money from the Australian people, so getting them fired from politics was the least I could do.

Maybe it was the money I took from their accounts that they are pissed about, I took the 10 million they embezzled and an extra million for my troubles. I moved the money they stole back to the people via a nice deposit in the name of Robin Hood (I liked the pun) to the make a wish foundation. I think they will make better use of it then those idiots would have, of that, I am definitely certain. Now I think about it, it could be any number of things really, I definitely haven't been the most law-abiding citizens lately. Although the several Robin Hood moments with stealing from the rich and giving to charities (as well as keeping a little of the rewards for myself of course) is my way of keeping my ledger in line. A way you can say for me to feel that I am of good conscience and for the most part I am pretty good with what I have done.

Tick, tick, tick that bloody clock is driving me crazy, how much longer do I need to sit here why they rummage through my things. Grrrr, this is seriously frustrating, it would just be simpler if I could just tell them that they aren't going to find a thing. This isn't where I work, it's not even in the same town but we will just have to let the theatrics continue. I will definitely need to keep a low profile for a while now until I can slip away and take a flight out of here. I already have access to the feds systems, so it will be easy to swap out my details for another's but I will have to wait for a few months before the details are a little fuzzy in the minds of anyone in this room.

It's strange what thoughts flow through your mind when you are completely fixed in a situation like this with no way of entertaining yourself. It's good to disconnect though, bounce around in my head for a while. I have been mulling over how it is I got here like really got here. What was the choice I made, and I think I may have narrowed it down? That choice that set me on the path to the dark side, on the criminal path instead of a reputable career. I could have been one of these AFP stooges walking around doing this to someone else, now that is a weird thought.

It was around October 1999, I was in high school and studying to do my HSC exams. I was truly crashing here, I was starting to drown in all this pressure and I could see that my friends were doing the same. The pressure of succeeding was horrible, I was never any good at maths or English, electronics was definitely my thing. That I had always known but if I failed these exams I would not get into any university, I would end up working at the potato farm or worse the local abattoirs for the rest of my life. I was not going to let that happen, that was for sure.

So, I am sure you can guess what happened next, I got the papers, all of them. It was actually pretty easy the staff who looked after the tests bragged about the fact that they controlled the papers until they would be sent out to students, they also used pretty weak passwords on all of their accounts. I could have had some real fun with their social media, but it was about my future and that meant I had to get those exam papers ahead of time. It was my only way I could be prepared.

I passed those tests, with pretty good marks, not great though as to raise attention to me but way better than I could have done without the cheat sheets so to speak. I do feel that that was my turning point though, as after that one choice to steal the exam papers, I kept using my abilities to change things that didn't align with what I thought I deserved or thought I needed. It was easy, it really was and after a while It wasn't even a challenge for me anymore. So, I started to get a bit more brazen and pushed the limits. I guess that is how I ended up here running my perfect Sunday roast, I was looking forward to eating pork crackling. Typical though I finally nail the crackle and it is sitting on the bench going cold. My mouth is starting to water just thinking about it but sadly I don't think I will be getting something to eat for a while, these guys look like they are just starting to settle in.

I am confident they aren't going to find anything that can really pin anything to me for any of the many things I have done, I have been careful about my whole life, I never mix my extracurricular activities with my daily life. Unless they have found my bunker than they are going to have to cut me loose at some point. Even if they don't it won't matter in another 24 hrs, the systems are designed to self-erase and burn themselves out if I don't login in and reset the counters every week and today is day six. Maybe they have found the bunker, Nah that isn't possible, I have been careful. I guess it doesn't matter even if they do, I couldn't even crack my encryption, so I am sure they won't even get close without setting off some kind of defences I have in place which will destroy any chance of them getting anything.

I wonder if they really know who they are dealing with, do they know what I am actually capable of? It's doubtful, my closest friends don't even know about my double life, my hidden persona who hunts for victims at night, if they are a leech on society or just someone that needs to learn a lesson, then I teach them with the only way it hurts. I take their money, I publicly display to the world how they really are and not the rubbish they pretend to be. The real person under the bull crap they try to have everyone believe. It's a public service really. So, what if I make a good cut from the transactions, hey I need tools to work, anonymity costs money you know and lots of it.

I know I am not the good guy in this story, I profit from bringing down these leeches, it's charities and the volunteers for these organisations who are the good ones. The donations are my way of feeling a little better about this and the targets are always people that deserve to be brought to light. Maybe it is my turn, maybe this is the end of the line for me, maybe I have finally been caught. Haha, Nah I really don't think any of these people are smart enough to have caught me out. I guess I will just have to sit here in my own thoughts waiting to see what it is they think they know about me.

Tick, tick, tick Grrrr. I swear if they let me go I am going to smash the hell out of that clock. It's like a sledgehammer just smashing away at my skull. This is really going to be a long night that is for sure. "Hey, you in the glasses. How about cutting me a few slices of pork over there?". Wow, I just got a look that clearly indicates I am not getting any of that food tonight. I might as well just admit to myself I am not going anywhere for now, so I may as well get some sleep. I am sure they will wake me up when I start to snore, or they are ready to go. Either way, I think it's time for some shut eye and for me to get the hell out of my head.

Chapter 50 - The Cyber Spook - A hacker fantasy tale

Some days I wonder what it would have been like if I hadn't met David. I met him in my final year of high school when the Australian defence force put together a program that would let budding cyber whiz kids test their skills against some of the country's best and brightest in cyber offence and defence tactics. It was a new program that was in its first year, they weren't sure how well or bad it was going to be, but they needed a way to bring in some new talent. They had never met someone like me before though. Ever since I was a little girl I just had a gift, It was almost as though I could feel machines, as though they were alive, and I just knew what I needed to do to get them to do what I wanted. It was really hard to explain how it worked but I could just make them do what I wanted.

If I wanted to gain access to any system I would just close my eyes, visualise the result and it was as though I could read the systems like a book and I could manipulate it as easy as moving through water. It was just like it was in my blood and I was meant to do this. When David came to the school he was presenting himself as a major in the Australian defence force but I later found out that this was not quite the truth, he was actually part of a hidden division in the Australian defence portfolio the ASD (Australian Signals Directorate) and it was, in fact, a clandestine organisation that was responsible for taking out by any means necessary the cyber assets of our enemies. This unit was the elite in cyber warfare and it was required to stay ahead of its enemies without showing that it even existed. Defend and attack from the shadows.

I obviously didn't know that at the time and I had never really shown anyone the skills that I had, I didn't even tell my parents what I could do. When I saw the flyer about the 1st prize of a state-of-the-art computer system of the winner's choice to a value of up to $10K. The winner would also get a full ride scholarship to the University of Canberra. I was really tempted, that kind of system was completely out of my reach and I could almost feel the rush that using it would give, the power and precise manoeuvres I could conduct would be almost unstoppable. I was hooked I had to be part of it.

I wasn't sure if I wanted to go to university and Canberra was a long way from home but maybe it would make my parents proud if I was the first in my family to attend university. It was decided I would sign up and at least check out what it is they wanted to do. When the day actually came around my school was abuzz with excitement and at least 100 students had signed up for the event. They went through the usual fluff show with all the talk about finding the best of the best to help make Australia secure. I will admit I had glazed over for most of that part of the show but snapped to focus as soon as they announced that it was time for contestants to connect to the wifi network they had set up for the event and see what it is they could find for the first level of the contest.

I was admittedly a bit disappointed with it, they had just set up a set of vulnerable virtual machines with some flags to find. I was through all 10 machines in under 30 minutes. I wrote down my answers and packed up my laptop. I put my answer sheet on the main desk up against the front of the event and started to walk out. That is when David saw me, he picked up my answer sheet and I could see a smile on his face. Suddenly he ushered for one of his men to come to him, he leaned in and whispered something in his ear. Within seconds I had four men surround me, "Miss please follow us".

They ushered me through into a room at the back of the hall, David appeared a few moments later and had a look of intrigue almost on his face. "Tell me, something kid, how did you gain control of all these systems so fast?" I didn't answer just looked at him for a moment, he held my gaze and after a few more minutes he handed me a card. "I will send a car for you at your home tomorrow, 0900 hours. Be ready. " I honestly didn't know what was going to happen now, but the men cleared the room as he did and left me to my own devices.

Later that night I was laying in bed thinking about the events of the day and I must admit I had absolutely no idea what I was in for the following day, I was so naïve. I really didn't sleep much that night, my brain was analysing all possible scenarios, what was it that was going to happen, would I be heading to lockup or was I the winner of yesterday's sideshow. I don't think that show was real though, I think it was all just smoke and mirrors to bring out hackers from hiding. I guess it worked, I showed myself to them and blew away their tests in minutes not hours. I was still disappointed that it was so easy to break open the systems, I really thought that It could have been a challenge for me, but it wasn't. I later found out that I was the only one to get through all machines and the closest one to me took 6 hours to break through the first 6. Maybe it was harder than I thought.

As was directed a black saloon car arrived at 0900 and a driver took me to a site about two hours outside of town. I didn't even know this place existed and I had lived in Brisbane all my life. After sitting around in a guarded room (some interesting guys in suits were watching me like I was going to steal the last doughnut in the box or something) for about an hour David appeared through a door on my left and sat down in front of me. He tossed a folder on the table and said, "Take a look". What was this all about, I didn't know what was going on.

The folder contained information on a computer system, user names and details that all seem to point back towards one cyber persona – StormySea. David waved towards one of his men and they brought forward a laptop, he put it in front of me and all he said was "find who that is? I want to know everything about them. You have 1 hour.

I spent a couple of minutes looking over the machine they gave me, it was a beast I really liked it and it was preloaded with almost anything I could think of that I may need. What should I do, I really wasn't sure what I should do but I just decided what the heck, let's show these boys what I can do. I closed my eyes and visualised my target, then when I was ready, I got to work. It was a few minutes out from my hour deadline but I decided to have a little fun, so I got myself access to everything I could find that belonged to David, just for giggles you know. Probably not my smartest move but definitely a good way I thought to lighten the mood a little.

I was playing around making some changes to David's email signature when he returned, I just hit send to an all staff email from him before I looked up to meet his gaze. I wasn't sure if that was a mistake or not but hey it was done. "So, what do you have for me?" I held his gaze for a moment then decided to play ball, I looked down at the laptop and hit send on the email I had drafted to him. His phone vibrated suddenly and kept doing it for almost a full minute. I said, "You should probably check that." He looked at me for a few moments and then did as I told him.

He looked a little confused at first then he had a realisation I had owned his accounts but not only had I taken over all of his accounts, but I also had everything he could ever want to know about the guy who went as the persona StormySea based out of Ukraine. I even had access information for all of his online storage and virtual platforms. They could access everything this guy had online, and I had also given him information on three other accounts that are affiliated with him and all of their online account info as well. Everything was just there for him.

He waved over one of his men again and they took the equipment off me. They all walked out of the room and I was left to my own devices for several hours. Someone brought in food and drink. I had an escort to the bathroom a couple of times when I asked. I never saw David again that day, but after a few hours I was driven back home, and the driver just left without saying anything further to me. Maybe the attack on his accounts was a little too much and didn't go down well.

Weeks went by and I hadn't heard anything more about it. No fancy new computer for me I guess. That's what I get for being a smart ass right. When I got home from school however, I saw three black cars parked in front of my house. David was inside talking to my parents, they all paused when I entered the room. My father looks at me "We are told that you are very good at computers and you have won some sort of competition at school?" I nodded and started to put my books down on the table. "He also indicated that you have an opportunity to attend the University of Canberra or do a direct traineeship with the defence force division the major leads", I was a little surprised but nodded again.

David turns to my parents and asks if they can give us a few minutes, they agreed and left us alone in the dining room. "You really made a mess of my accounts, you know that? I am also still getting replies about the date invitations you sent out to all my contacts as well, it would appear that I am a popular man". I was having trouble holding back the smile at that one. He talked about what I had done for what seemed like an hour going over everything including the fact I had been able to get more information on the StormySea target than any of his team in an hour (less obviously as I hacked him as well), it took them weeks to verify it all and they used my information to bring him down.

"You have a job with us if you want it. You will not find anywhere else but the ASD where you will have a licence to hack". That day was the day I became a spook, I was a secret member of a team that hacked anyone or anything that threatened our country or who were just downright horrible people. No one knows what I do outside of my team though, my loved ones think I am just an IT girl for the army, I guess they are a little right.

www.ingramcontent.com/pod-product-compliance
Lightning Source LLC
LaVergne TN
LVHW050150060326
832904LV00003B/88